2349 Computer and
Space Sciences Building
University of Maryland
College Park, Maryland 20742-2461

Thinking Through

Mathematics

**Fostering Inquiry
and Communication
in Mathematics
Classrooms**

Edward A. Silver

Jeremy Kilpatrick

Beth Schlesinger

Dennie Palmer Wolf, *Coordinating Editor*

Robert Orrill, *Executive Editor*

College Entrance Examination Board, New York, 1990

Edward A. Silver is Professor of Cognitive Studies and Mathematics Education in the School of Education and Senior Scientist at the Learning Research and Development Center at the University of Pittsburgh, Pennsylvania.

Jeremy Kilpatrick is Professor of Mathematics Education at the University of Georgia, Athens, Georgia.

Beth Schlesinger teaches mathematics at San Diego High School in the San Diego Unified School District, California.

Dennie Palmer Wolf is Senior Research Associate, Project Zero, Harvard School of Education, Cambridge, Massachusetts.

Robert Orrill is Associate Director for Academic Development, Office of Academic Affairs, The College Board, New York.

For Rebecca and Monica—EAS

For Barton and Judson—JK

For Robert and Steven—BS

Researchers are encouraged to express freely their professional judgment. Therefore, points of view or opinions stated in College Board books do not necessarily represent official College Board position or policy.

Single copies of *Thinking Through Mathematics* can be purchased for $8.95. Orders for five or more copies receive a 20 percent discount. Payment or purchase order should be addressed to: College Board Publications, Box 886, New York, New York 10101-0886.

Copyright © 1990 by College Entrance Examination Board. All rights reserved. The College Board, the EQ design, and the acorn logo are registered trademarks of the College Entrance Examination Board. The Educational EQuality Project and EQuality are trademarks owned by the College Entrance Examination Board.

Cover photograph: Judith D. Sedwick

Library of Congress Catalog Number: 90-080565

ISBN: 0-87447-382-9

Printed in the United States of America

9 8 7 6 5 4 3 2

Contents

Foreword .... *v*

Acknowledgments .... *vii*

Preface .... *ix*

1. Rethinking Mathematics Teaching *1*
   The Changing View of Mathematics Education: Beyond Bits
   and Hierarchies *5*
      Constructing Understanding *6*
      The Social Side of Mathematics *7*
   An Emerging View of Mathematical Thinking *8*
      Applying Mathematics *9*
      The Power of Open-Ended Problems *10*
      Creating Rather than Answering Problems *10*
      The Place of Conjecture: The Provisional Side of
         Mathematics *12*

**2. Communicating Mathematical Ideas: Private and
    Public Discourse** *15*
   Opening Windows: Teachers Learning from Students *16*
   Getting It Straight: Students Learning from Themselves *21*
   A Community of Learners: Students Learning from Each Other .. *22*

**3. Transforming Ordinary Situations Into
    Extraordinary Opportunities** *25*
   Opportunities for Mathematical Speculation *25*
   Opportunities for Thoughtful Mathematical Analysis *30*
   Opportunities for Convincing Oneself and Others *38*

4. Getting Started *41*
   Providing Opportunities: Opening Up Lines of Inquiry and
      Communication *41*
   Timely Advice *42*
      On Changing the Rules: Expect Some Resistance *42*
      Rome Wasn't Built in a Day: Make Time and Have Patience .. *43*
      Setting Modest Goals: The 10 Percent Solution *43*
   Possible Entry Points *44*
   A Case in Point: How Mrs. Holmes Got Started *45*
      Recovering from the Opening Fiasco: A Strategic Approach .. *46*

On the Dotted Line: Mrs. Holmes Helps Her Students
   Write About Mathematics........................ *49*
Nothing Succeeds Like Success *50*
Looking Back: Mrs. Holmes Reflects on the Year *51*
Lessons to Be Learned: Further Reflections on
   Mrs. Holmes's Experience....................... *51*

References.. *53*

Figures

Figure 1-1. "Smooth" Graph of Tree-Replacement Problem *3*
Figure 1-2. "Unconnected" Graph of Tree-Replacement Problem *4*
Figure 2-1. José's Algebraic Approach to the Chicken-and-Rabbit
   Problem *17*
Figure 2-2. Lora's Visual Approach to the Chicken-and-Rabbit
   Problem *18*
Figure 2-3. Robert's Algebraic Approach to the Chicken-and-Rabbit
   Problem *18*
Figure 2-4. Karen's Arithmetic Approach to the Chicken-and-Rabbit
   Problem *19*
Figure 2-5. Responses to "Tell me everything you know about
   circles" *20*
Figure 3-1. Examples of Open-Ended Problems Used by Mr. Ruiz.... *26*
Figure 3-2. Graph of the Year and Difference for Mrs. Herrera's Age
   Problem *32*
Figure 3-3. Graph of the Year and Ratio for Mrs. Herrera's
   Age Problem *33*
Figure 3-4. Angle-Bisector Problem Used by Mr. Jones *34*
Figure 3-5. Kerwin's and John's Solution to the Angle-Bisector
   Problem *35*
Figure 3-6. Tracie's Solution to the Angle-Bisector Problem *36*
Figure 3-7. Ramona's and Melanie's Solution to the Angle-Bisector
   Problem *37*
Figure 4-1. Kuo's Solution to the Handshake Problem *47*
Figure 4-2. Melanie's Solution to the Handshake Problem *47*
Figure 4-3. Tyrone's Solution to the Handshake Problem *48*

Foreword

Thinking Through Mathematics is one of a new series of publications initiated by the College Board's Educational EQuality Project—a ten-year effort to improve the quality of secondary education and to ensure equal access to college for all students. The books in this series will address what many observers are beginning to recognize as the central problem in significant educational change: the work of teaching all students, not a few, how to become competent thinkers (Resnick and Klopfer 1989). This work, if successful, will bring into existence a new conception of the place of thinking in the high school curriculum. Rather than being treated as a special, separate, and final skill, thinking will become the substance of the most basic classroom activities for all students in all subject areas. This series is designed to convey through discussion and example how many teachers are already making this new conception a reality in their classrooms—how, that is, they make even ordinary moments occasions of thought for their students.

Thinking Through Mathematics focuses on how mathematics teaching and learning can be improved by developing more powerful approaches to connecting thinking and mathematics. In so doing, it explains changing perspectives on what it means to learn and do mathematics, and explores how these perspectives can be incorporated into the teaching of secondary school mathematics. Other books in the series will address thinking in history, science, the arts, English, and foreign language. Each will explore how immersion in knowledge is required for thinking. All of the books, however, draw on both recent cognitive research and actual instances of classroom practice to convey how thinking should not be an activity that comes late in the curriculum after the acquisition of content, but rather is integral to successful learning in even the most commonplace and basic classroom situations. With deliberate purpose, *Thinking Through Mathematics* and the other books in this series constitute a response to what Lauren Resnick has described as the urgent need "to take seriously the aspiration of making thinking . . . a regular part of a school program for all the population, even minorities, even non-English speakers, even the poor" (Resnick 1987).

Robert Orrill
Office of Academic Affairs
The College Board

Acknowledgments

We are grateful to many people for their assistance during the preparation of this book. To Mrs. Shirley Cleland and her students at San Diego High School, we extend our thanks for trying several of the activities. Many of the teacher participants in the San Diego Mathematics Project and the San Diego Mathematics Teacher Enhancement Project contributed to this volume by sharing with us their experiences working with some of the activities discussed here. We would also like to acknowledge the roots of this book in the College Board Educational EQuality Summer Institutes held in 1985 in Santa Fe, New Mexico, and in 1986-88 in Santa Cruz, California. Teachers and faculty in those institutes helped to raise issues and clarify various ideas about integrating the teaching of thinking with the teaching of mathematics. Thanks also go to the students in the Master's level research seminar at the University of Georgia for trying several of the writing activities with students in their classes, and to the students in the Master's level course on mathematical problem solving at the University of Pittsburgh for trying many of the open-ended problem-solving activities.

To Sandra Marshall, Director of the Center for Research in Mathematics and Science Education at San Diego State University, we extend our thanks for her generosity in providing us with a pleasant place to meet and plan our work during its formative stages.

Special thanks go to Margaret Smith for her perceptive comments on a nearly final draft and her considerable help during the final stages of production. We are also grateful to the members of the College Board Mathematics Advisory Committee for their encouragement and for their comments on an early draft of the manuscript, to Dennie Wolf for her helpful editorial suggestions, and to Bob Orrill for his patience and support throughout the process.

In addition to the support provided by the College Board, preparation of this book was also influenced by work supported by grants to the first author from the National Science Foundation (MDR-8850580) and from the Ford Foundation (890-0572). The views expressed herein are those of the authors and do not necessarily reflect the views of either the National Science Foundation or the Ford Foundation.

Edward A. Silver
Jeremy Kilpatrick
Beth Schlesinger

Preface

We hope this small book will stimulate discussion among teachers about the implications of some changing perspectives on what it means to learn and do mathematics, and how those perspectives can be incorporated into the teaching of secondary school mathematics. In Chapter 1, we offer a view of mathematics as emerging largely from individual and social activity rather than only from textbooks, worksheets, and tradition. We also picture the learner of mathematics as someone who actively constructs meaning rather than passively receives it. In Chapter 2, we consider how a greater emphasis on communication in the mathematics classroom yields dividends. Discussion and debate, recording, and writing stimulate and uncover students' learning and thinking and lead to deeper understanding by both teachers and students. Next, we explore how teachers might encourage greater inquiry and communication in a secondary school class by making minor, but thought-provoking, changes in ordinary problems and situations. Finally, we give some practical advice on getting started in the risky, but rewarding, business of transforming the mathematics classroom into a place where students are expected not only to absorb and consume mathematics but also to produce and think about it.

A central theme in this book is that mathematics teaching and learning will be improved if we can find more powerful approaches to connecting thinking and mathematics. Students can learn *thinking through mathematics* when they see the mathematics classroom as a place in which there is an open invitation to be thoughtful, and when they understand that mathematics is connected to a rich variety of interesting situations and problems. Toward that end, we have provided examples as a way of stimulating teachers, working in groups or individually, to find, construct, and invent their own activities and problem situations. Interesting problems and mathematically promising situations are all around us, once we become sensitive to finding them, become active in looking for them, and begin collecting them systematically.

We expect that some of our example problems will be familiar to many teachers, but that others will be new. In our discussion of the examples, we have tried to illustrate that interesting classroom opportunities for mathematical thinking can be constructed not only from the consideration of novel situations drawn from newspaper stories or everyday experience but also from the use or adaptation of well-known problems or tasks taken directly from textbooks.

If we are to involve our students and ourselves in a process of mathematical inquiry and investigation like the one outlined here, we will all need suggestions, encouragement, and support. We hope that the information contained here will provide the impetus for many of us to begin and sustain the journey. We also hope that those who find some measure of success will share their experiences with others in the mathematics education community. Our community will grow and thrive to the extent that we can nurture not only our students' mathematical thinking but also our own professional growth and that of our colleagues.

Making mathematics come to life through the consideration of interesting problems requires no special talent, and it is not a gift restricted to a few expert pedagogues. Teachers may find it useful to collaborate with colleagues in developing collections of problems and situations that have worked well for them, but ultimately each teacher has to find the ones that fit with his or her own vision of mathematics, teaching style, and instructional goals. We hope the examples in this book will provide a starting point for a teacher's reflection on teaching mathematics—a personal opportunity for *thinking through mathematics*.

1. Rethinking Mathematics Teaching

Each year New York City loses thousands of trees in its parks and along its streets to contractors who rip out trees to clear the view, and to vandals who chop down trees for kicks. The Parks Department is responsible for ensuring that the trees are restored by those who remove or destroy them. At one time, the department used a simple formula to determine how trees were to be replaced—a tree for a tree. That is, a tree that was, say, 12 inches in diameter could be replaced by a single tree of any diameter. Later, a second formula was used—an inch for an inch. A 12-inch tree was to be replaced by four 3-inch trees, or three 4-inch trees, or any other combination as long as the combined diameters were 12 inches. Today, a third formula is in use—a square inch for a square inch. That is, the combined area of the replacement tree trunks, as measured in a cross-section, must equal the area of the original tree trunk.

Mrs. Walsh, a mathematics teacher, read about these formulas in a newspaper article. She decided that the situation of tree replacement in New York City could serve as the basis for some class work on functions. She wanted a topic from real life that students would understand and in which they could explore mathematical ideas. She wanted her students to realize that doing mathematics was a common activity, and she wanted

Notes

them to view themselves as capable of making sense of new problem situations in the world around them. She describes her experience:

> I had been reading a lot of articles about the importance of developing thinking skills in mathematics. I wanted to find some problems that were different from the textbook. Other teachers had told me that the daily newspaper was a good source of realistic situations. I decided to ask the students in my algebra class to make up their own problems from situations in the paper. I thought that was better than giving them problems I had made up. I told them about the three tree-replacement functions that had been used in New York. We agreed to simplify the situation by first considering the replacement of just one 12-inch tree. The students started out working in groups on the problem of graphing the number of replacement trees as a function of the diameter of the trees used as replacements—assuming all had the same diameter. I asked each group to look only at replacement diameters between 0 and 12 inches and to construct the graph for each of the three rules. I also asked them to write a function rule for each set of points.

Mrs. Walsh used the graphing activity as a way of getting into the situation because the process of plotting points requires students to consider how the function behaves and can provide a visual and conceptual basis for an eventual algebraic formulation. The students could see the functional relationship corresponding to each rule. If students are going to make up their own problems to solve, they need some context out of which the problems can arise. The tree-replacement context is rich in mathematical relationships, but students need time to see and explore. Once they understand better what is going on, they are in a position to think about problems.

By asking her students to work in groups, Mrs. Walsh enabled them to divide up the graphing tasks—choosing values, calculating coordinates, and plotting points—thus making more time available for thinking about the relationships. They could also monitor each other's understanding.

> Some of the groups connected the points that they plotted to create a smooth graph, but others left the points unconnected (see Figures 1-1 and 1-2 below). We got into a very interesting discussion, actually a heated debate, about whether or not the points should be connected. Although the students didn't use these terms, the argument centered on the fact that the number of replacement trees was a discrete variable while the number of inches in the diameters was a continuous variable. Students who wanted to connect the points agreed that the number of replacement trees had to be a whole number, but they also knew the diameters did not have to be whole numbers. In the end, we all agreed that the points should not be connected because one of the variables was not continuous. That was one of the most interesting discussions about graphing that I've ever had with a class.

Several groups needed some help writing the second and third function rules algebraically. They could say in words what was needed, but they couldn't write it until we introduced the notation that x would be the diameter of the replacement trees and n would be the number of trees. That helped, and most were able to get $n = 12/x$ for the second. A few got $n = 144/x^2$ for the third. When I told them that they had been dealing with hyperbolas and other complicated functions, they were impressed!

Later, I asked the students to work together in the same groups to make up two problems about replacement trees. One problem was to be like the word problems in their textbook. I didn't give them any examples because I didn't want them to copy what I said; besides, they all know what those problems look like. But I asked that the second problem be more challenging and unusual. I suggested that they use their graphs to give them ideas. I told them that once their problems were written, the groups would exchange and try to solve each others' problems.

Notes

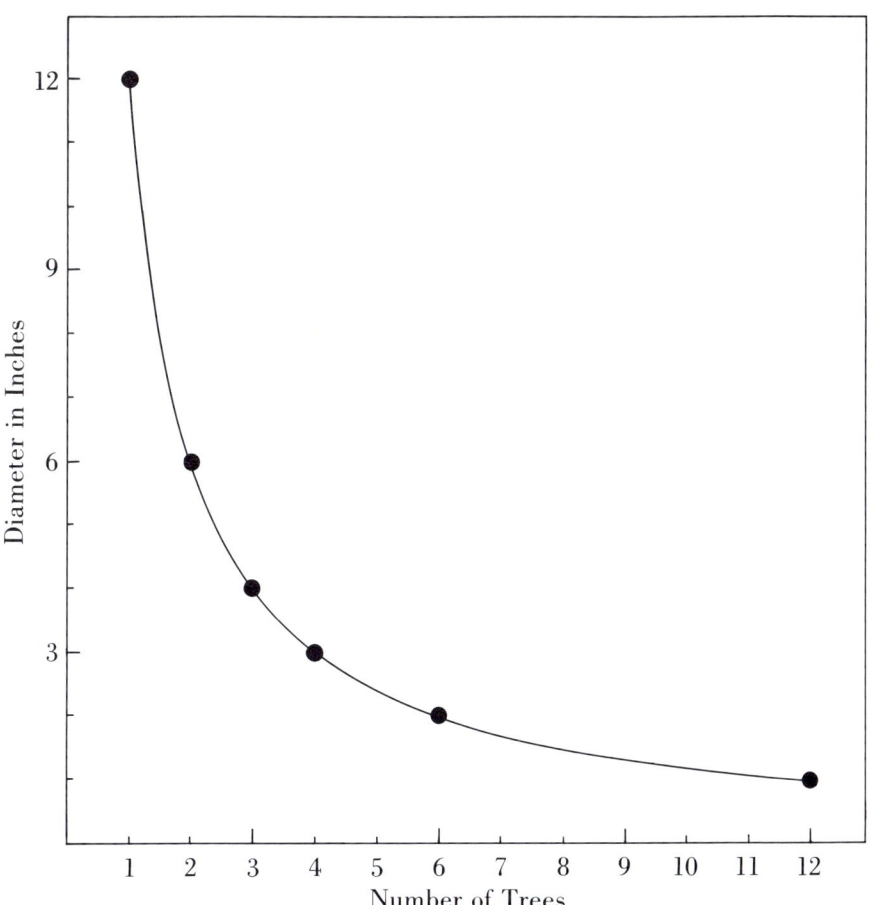

Figure 1-1. "Smooth" Graph of Tree-Replacement Problem.

Notes

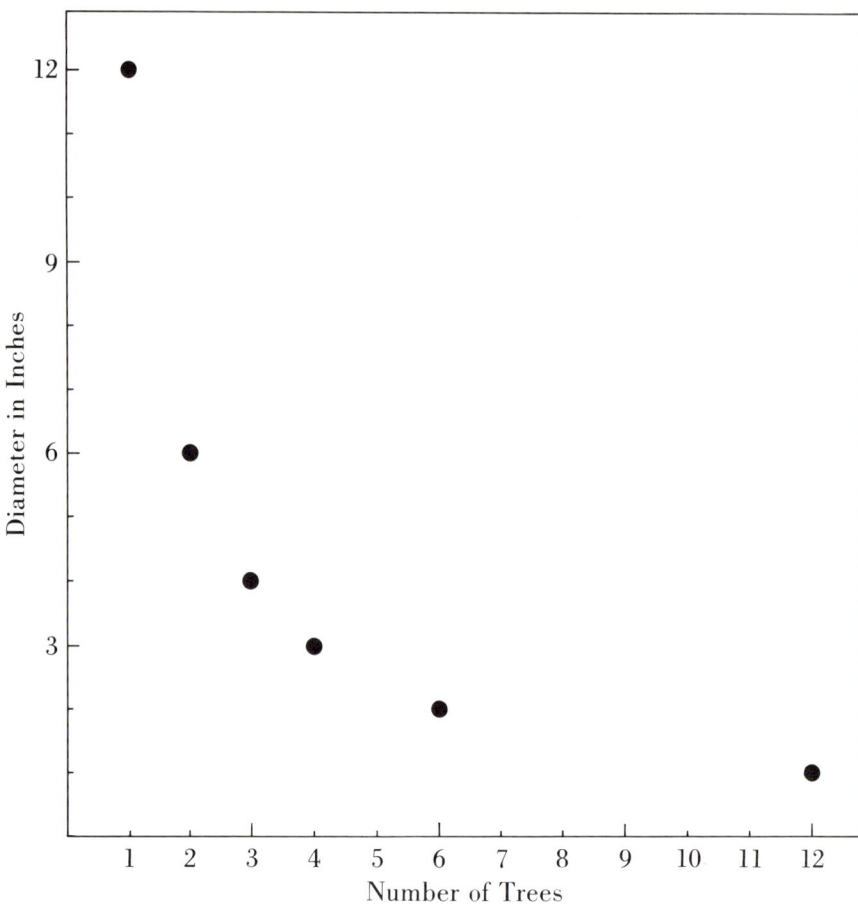

Figure 1-2. "Unconnected" Graph of Tree-Replacement Problem.

The teacher asked students to work in groups because she was concerned that although making up a textbook-like problem would be easy, making up a challenging and nonroutine problem might be rather difficult for students unless they had others to help them. The teacher wanted students to write two kinds of problems, partly because she wanted them to be successful on at least part of the task, but also because she wanted them to understand something of the difference between the two types of problems. She had been emphasizing techniques of problem solving, and she thought that by having the students construct their own problems she could help them improve their understanding of what a problem is, and what can make it difficult or easy.

> I found that while the students had worked quietly and diligently on the graphing part, they got very noisy when it came to writing problems. There were many disagreements about what a good problem would be. I had to keep moving from group to group to get them to lower their voices. Finally, I had to tell them that we would stop our group work for the time being and come back to this problem writing

4   THINKING THROUGH MATHEMATICS

another day. That did it, because they really did enjoy the activity. They just couldn't do it calmly enough.

Later that week, we got back into the same groups and began working again on our problems—this time, more successfully and quietly. Some students had been thinking about the graphs and what they meant. None of the groups had any real trouble writing a textbook word problem. Almost everyone came up with "How many more 3-inch trees than 4-inch trees are needed to replace a 12-inch tree under the square-inch rule?"

But they were very inventive with the nonroutine problems. One group made up a story about a new Parks Superintendent who enforced a cubic-inch-for-a-cubic-inch rule. They wanted to know what difference that would make, compared to the square-inch rule, if a tree's height increased by a foot every time its diameter increased by an inch—not a very realistic assumption, but one that led to an interesting investigation. Another group asked how the increase from the first to the second formula compared with the increase from the second to the third formula. The other students had a lot of trouble solving that one algebraically, although they could see it geometrically. Some groups wrote problems that I myself couldn't see how to work, and some of their problems didn't seem to have any good solution. But the students obviously enjoyed inventing difficult problems and trying to work each others' problems. In the process, they may have learned something about what makes a good problem. If nothing else, they learned that challenging problems are not easy to write.

The students in Mrs. Walsh's class have learned something about functions in a realistic context. The Parks Department wanted to increase the amount of wood replaced when a tree is destroyed. Each functional relationship the department came up with was more complicated than the preceding one, but yielded a higher return. The students saw this graphically, and they used their knowledge of algebra to express each relationship. They saw how algebra could be applied to explore and understand a complex, practical situation.

The Changing View of Mathematics Education: Beyond Bits and Hierarchies

Mrs. Walsh's efforts to make algebra come alive for her students reflect changing views about what it means to learn and do mathematics. For years, teachers have complained that students are bored by high school mathematics. By the time students graduate, they have forgotten most of what they were supposed to learn, and they proclaim, often with a hint of misplaced pride, "I never was any good at mathematics." The mathematics they say they were no good at was, by and large, the mathematics of routine computation and manipulation (usually with some formal proof

Notes

thrown in). Mathematics was little more than a collection of recipes that students were expected to memorize and practice.

Courses for teachers used to stress how important it is to break the teaching of a subject like mathematics into discrete pieces organized in a logical hierarchy, like the bricks in the walls of a building. The student begins at the bottom, with the simplest possible tasks, and works his or her way up the wall—the complex task having as prerequisites all the simpler tasks below it. Unfortunately, this metaphor for organizing instruction has two flaws: it ignores how people learn, and it distorts what they are learning.

The brick-wall metaphor is a helpful way of showing how a complex task can be broken down into simpler ones, and it can also help us to analyze apparently simple tasks to see how complex they really are. It reminds us about how the various things students are asked to do should fit together into some coherent whole. But many decades of research on human learning of complex subjects suggests that people do not always learn things bit by bit from the ground up (Resnick, 1987). They often jump into any situation with some knowledge, however rudimentary or inaccurate, and, even before they have mastered specific techniques, they begin fitting their knowledge into a larger picture. Students bring their own interpretations of tasks and concepts to the instructional process. Forcing them to master all the so-called prerequisites in a hierarchy before moving on dooms many of them to trivialized, repetitious instruction and keeps them from seeing where they are headed. They cannot see the building for the bricks.

More important, the metaphor assumes that everything worth considering about mathematics as a school subject can be put into hierarchically organized collections of "bits." But mathematics is more than sets of procedures to be mastered or rules to be memorized. It includes, among other things, the investigation of patterns, the formulation of hunches about what generalization might fit a pattern, the communication of arguments for or against various conjectures, and the understanding of fundamental concepts well enough to judge when to use a particular skill. It is as much about processes like problem solving, problem posing, conjecturing, justifying, and convincing as it is about the discrete bits that fit neatly into knowledge hierarchies (National Council of Teachers of Mathematics, 1989a). The logical structure implicit in much of mathematical knowledge has seduced us into believing that the hierarchical, simple-to-complex ordering that can be imposed on mathematics is synonymous with mathematics itself. When we stop to analyze what doing mathematics is really like, we see that the metaphor of bits or bricks does not fit.

Constructing Understanding

Mathematics teachers are realizing that learning is not receiving and remembering a transmitted message. Instead, "educational research offers compelling evidence that students learn mathematics well only when they *construct* their own mathematical understanding" (Mathematical Sciences Education Board, 1989, p. 58). This view is not new, but it is becoming more useful to teachers as its implications are being developed by researchers in cognitive psychology and mathematics education.

Constructing mathematics involves more than acquiring new concepts. It also involves *re-*constructing prior knowledge. Kieran's (1981) work on the difficulty students have in first learning to solve linear equations gives an example of how students' poor conceptions influence their acquisition of new mathematical knowledge and may require some instructional modifications. She found that, probably as a direct result of their extensive arithmetic experience, students beginning the study of algebra tend to think of the equal sign as a "do-something" signal. That is, one writes the equal sign to signal that it is time to perform some calculation indicated to the left of the equal sign, after which one writes the "answer"—the result of the calculation just performed—to the right of the sign. Students understand, more or less, what an equation like $3x + 5 = 17$ means because there is a single term on the right—the result of doing something. But, correspondingly, they have difficulty solving an equation like $3x + 5 = x + 17$. It seems unfinished, since the right side still has an addition to be performed. Kieran developed a sequence of activities that help students reorganize their understanding so that they could see the equal sign as a relational symbol, rather than as a signal to do something. She argued that teachers need to take seriously the conceptions that students bring to the algebra class from their experience in arithmetic, even when those conceptions are partial or limited.

When learning is seen as knowledge construction and reorganization, teachers can consider the special ways each student learns, and they can begin to view instruction not as the piling up of little bricks of knowledge but as an effort to help students make sense out of their world. Students are not blotters, absorbing experience in the shape presented to them, but active minds making meaning out of experience by constantly reconstructing and reorganizing their knowledge.

The Social Side of Mathematics

Knowledge—including mathematical knowledge—is now being seen as socially constructed. We mathematics teachers have commonly believed that, unlike teachers of English or social studies, we did not have to worry about promoting social interaction in our classroom. Oh, yes, we understood that mathematics is a language, but we tended not to see it as a language that needed to be developed by students actively communicating with one another and with us. Mathematics, we argued, was a language that one could learn alone.

Today, research and experience have taught us that mathematics is learned through a process of communication. Students need opportunities not just to listen, but to speak mathematics themselves—to discuss what they have observed, why procedures appear to work, or why they think

Notes

Notes

their solution is correct. Students read hundreds of mathematical exercises in their textbooks; they also need opportunities to read about mathematics and to write about the mathematical ideas they have.

The history of mathematics teaches us that communication and social interaction have played fundamental roles in the development of mathematical ideas. But students come away from high school mathematics unaware that mathematics has been constructed over time by a community of people with evolving ideas as to what is important and what should count as mathematics. Most students view mathematics as outside history. It emerged fully formed from Euclid's brow and has remained fixed ever since. The idea that mathematics is culturally determined, that it reflects currently held values, that it is beset with controversies about fundamentals and priorities is seldom considered in school. Yet there is growing evidence (e.g., Bishop, 1988) that attention to the social dimension of mathematics can enrich both the teaching and the learning of the subject. Students learn that there are contexts for the abstractions of school mathematics, and that there is a humanistic side to a subject that has seemed so devoid of humanity.

Mathematics is not the only thing students learn from the social world of the mathematics classroom. They also learn ways of behaving. They learn what it is that their society thinks is important for them to know. They learn what values their peers and their teacher place on mathematical ability, on verbal facility, on competition, on cooperation, on hard work, and on getting by. They learn whether to try, or to appear to try, or to ignore the teacher altogether. Classrooms are communities where people agree to behave in certain ways, and where they carry on an extended dialogue even when only a few are talking.

A mathematics course is like a conversation that is periodically interrupted, but that is gradually shaped by the participants into something they mutually share. Many teachers today believe that their mathematics courses will be more successful if they are organized so that students play a more active role, if the level of thinking is higher, and if the mathematics studied is more clearly connected to a sensible context.

An Emerging View of Mathematical Thinking

If students are to develop the power to use mathematics productively once they leave school, they need opportunities to use it productively while they are there. More attention needs to be focused on the thinking that students are doing about the mathematics they are learning. How are students encountering mathematics? How are they using it to solve problems? To reason? To communicate?

> To understand what they learn, [students] must enact for themselves verbs that permeate the mathematics curriculum: "examine," "represent," "transform," "solve," "apply," "prove," "communicate." This happens most readily when students work in groups, engage in

discussion, make presentations, and in other ways take charge of their own learning" (Mathematical Sciences Education Board, 1989, pp. 58-59).

Applying Mathematics

Most of us were taught that the most elegant mathematical thinking was "pure"—the power of sheer deduction and proof. This, too, is changing.

> Because mathematics is a foundation discipline for other disciplines and grows in direct proportion to its utility, we believe that the curriculum for all students must provide opportunities to develop an understanding of mathematical models, structures, and simulations applicable to many disciplines (National Council of Teachers of Mathematics, 1989a, p. 7).

Mathematics is a far different subject than it was when any of us were in school, primarily because of changes wrought by the computer.

> In recent years, computers have amplified the impact of applications; together, computers and applications have swept like a cyclone across the terrain of mathematics. Forces unleashed by the interaction of these intellectual storms have changed forever—and for the better—the morphology of mathematics. (Steen, 1988, p. 611).

Little of this change has been reflected in the school mathematics curriculum. Mathematics has become "the invisible culture of our age" (Mathematical Sciences Education Board, 1989, p. 32), and yet much of that culture—the knowledge people need to understand issues of public policy, the confidence they need to use mathematics imaginatively as a tool, the appreciation they might have for the recreational and aesthetic aspects of mathematics—is missing in schools today. Applications of mathematics are often neglected by teachers because they are not familiar with how the mathematics they are teaching might be applied, or because they do not want to spend valuable class time on activities seen as "outside mathematics." Students who develop mathematical models of practical situations, however, gain valuable experience in putting mathematics to use.

> Applications typically begin with an ill-defined situation outside mathematics—in economics, physics, engineering, biology, or almost any field of human activity. The job is to understand this situation as well as possible. The procedure is to make a mathematical model which we hope will shed some light on the situation we are trying to understand. Thus, the heart of applied mathematics is the injunction "Here is a situation; think about it" (Pollak, 1970, p. 328).

Students come to see the relevance of the mathematics they are learning, and when they do, it can be a powerful force in motivating the further study of mathematics.

Notes

Notes

Many mathematics teachers, like Mrs. Walsh, develop, collect, and cultivate a set of practical situations in which any of several previously learned concepts and techniques might be applied. Such strategies erode the "phoniness" of students being presented with an application problem right after studying the mathematics they are expected to use in identifying, modeling, and solving it.

To return to the opening example, teachers need not have just completed a unit on functions in order to help students apply mathematics to the situation of tree replacement in New York City. They could begin by presenting the original formula—a tree for a tree—and ask the students to come up with other formulas and consider what their effects might be. Alternative formulas could be graphed and discussed. If a quadratic formula relating number and diameter were proposed and the students had not studied quadratic equations, some sort of graph could still be plotted. Further investigation of the formula could always be postponed. But the goal of formulating the problem, identifying the important and unimportant variables, developing a mathematical model, and using the model to make inferences relevant to solving the problem would still have been met.

The Power of Open-Ended Problems

What other formulas could the Parks Department use? is the sort of problem that seldom appears in mathematics instruction. Not only because it deals with a real situation, but also because it is open ended. In general, open-ended problems, even those that are strictly mathematical in nature, tend not to be found in school mathematics. *What properties do prime numbers have?* is an open-ended problem that students can explore, but that also drives the work of some research mathematicians. An advantage of open-ended problems is that, when they are used with groups of students, teachers usually do not need to worry about getting students started. Anyone who understands what the problem is asking can make a start, however tentative. Students working together invariably find something in the problem that they can work on. Sometimes they narrow the question; sometimes they refine it. Often, they choose one idea in the problem and pursue that. Groups that get stuck in one line of investigation can be asked to go back to the original problem and think of it another way. Although open-ended problems may be more difficult for a teacher to manage because the students may go off in unexpected directions, teachers who have used such problems in their classes say that students enjoy the challenge they provide. Mathematics class then becomes a setting for exploration and originality.

Creating Rather than Answering Problems

Students need practice in formulating mathematical problems for themselves. If they are always presented with well-formulated problems that contain just the information needed for a solution, how can they learn to deal with situations in which appropriate mathematical ideas and techniques are not obvious—that is, situations in real life?

Students can be helped to develop problem-formulating skills by giving them an ill-formulated or a partially formulated problem and asking them to restate it. Mr. Kaminski, a ninth-grade general mathematics teacher in a large urban high school, has an activity he calls "Fix It" that he often uses on Fridays or whenever there is some time at the end of a class period. Groups of students compete to see which group can come up with the best problem (and solution), given an incomplete problem. The problems and solutions are written on large sheets of paper and posted on the bulletin board. Students in his other general mathematics classes get to vote on which one they like best. Here he illustrates the wide range of mathematical thinking open-ended problems can provoke:

> I gave this Fix-It Problem to the students in my fifth-period class:
>
> *Tammy bought a record for $4.95 and paid 7 percent sales tax.*
>
> Almost all the students saw right away that the question was missing. One group just asked the obvious question of what the amount was that she paid. Another tried to complicate the problem by asking what change she got from a $10 bill. The group that won wrote this:
>
> *Tammy bought a record for $4.95. She paid $5.30. What is the sales-tax rate?*
>
> In writing up their solution, the group said that the tax of 35 cents would probably be more than the actual amount, and therefore the tax rate would probably be less than 7.07 percent. They said the most reasonable answer was 7 percent. I thought that showed a lot of insight into how rounding off works in this case. Students who voted for the problem said they liked it because the solution was "realistic."

Even teachers who are not especially interested in problem creation as a goal of mathematics instruction have found that having students write their own problems and change existing problems can be a valuable teaching technique. It helps students see how problems are constructed. It encourages students to take an active part in changing a problem into something they can solve. It gives them the feeling of competence and mastery. For the teacher, it may illuminate some of the difficulties students can have in understanding a problem situation.

Problem posing is almost always overlooked in discussions of the importance of problem solving in the curriculum. Mathematics textbooks have an especially difficult time with problem posing because it is so open ended. Nonetheless, it ought to be given the same emphasis in instruction that problem solving is beginning to receive. It makes use of thinking abilities that need to be developed, both for their own sake and for their value in developing other abilities.

Notes

Notes

The Place of Conjecture: The Provisional Side of Mathematics

Although many people believe that deductive proof helps define the essence of mathematical reasoning, proofs have been disappearing over the past decade or so from their traditional home, the course in Euclidean geometry. There is no denying that pupils find the formal two-column proof difficult to learn, and that teachers often find it frustrating to teach. Courses are being restructured so that geometry without proof has become an option for many students, even those going on to college. As proofs are being de-emphasized in geometry, we have an opportunity to give more attention to other important mathematical processes like conjecture. When students are given the opportunity to make mathematical conjectures, they come in contact with mathematics in the making, with mathematics as it is practiced. Students who have made a mathematical conjecture have invested some of themselves in the enterprise of doing mathematics. Making a proof may be too difficult for some students, but making a conjecture, and then engaging in some reasoning about that conjecture, is something any student can do. In courses such as history and science, students learn what counts as evidence and how an argument can be developed to advance a claim. They need to learn that in mathematics, too, one can work out a position and gather evidence to support it.

Because of their previous experiences with mathematics, many students are afraid to make guesses or propose conjectures in class. They are understandably reluctant to expose their ignorance and confusion to public view. Mathematics has been for them a subject in which you are either right or wrong, and there is too much at risk in being wrong. If teachers want their students to grow in their ability to make conjectures about mathematical ideas, they may need to work hard to overcome the negative effects of previous experience. They need to make their classroom a place where informed guessing is encouraged.

Students need to know that it is only by making a good guess, probing and improving it, and supporting it with evidence, that anyone can do mathematics at all. Anything else is just memorizing. Mathematics in the making requires a willingness to take risks by offering a guess. Of course, as George Polya often remarked, the guess should not be a wild guess; anyone can make a wild guess. The person who is thinking mathematically, however, will offer a conjecture that has some basis in the situation at hand. Then the guesser needs to have the courage to test the guess and, if necessary, to replace it with a better one. If a deductive argument can be given to support the conjecture, fine. But if such an argument cannot be found, inductive evidence can make the conjecture more plausible, even though it cannot provide a finished proof (Polya, 1981).

Firsthand experience in looking for patterns in data, formulating conjectures about those patterns, and exploring the consequences of the conjectures can help students see the exploratory, provisional side of mathematics that is usually hidden from view in school. Statistical data such as those provided by the daily newspaper in the sports results, the stock-market quotations, and the weather report can be used as a source of conjectures about trends and relationships. Students can make tables that relate numbers to their divisors, or the perimeters of polygons to their areas, in order to generate data for analysis. The teacher should recognize that the exploration of data in search of patterns takes time, but that students who do not excel in other mathematical activities may flourish here. Exploring patterns is a legitimate part of mathematics.

> Students should be encouraged to observe and describe all sorts of patterns in the world around them: plowed fields, haystacks, architecture, paintings, leaves on trees, spirals on pineapples, and so on. As the students mature, instructional efforts can move toward building a firm grasp of the interplay among tables of data, graphs, and algebraic expressions as ways of describing functions and solving problems (National Council of Teachers of Mathematics, 1989a, pp. 98-99).

Teachers should neither be embarrassed about encouraging pattern-finding activities nor reluctant to include them in instruction.

The ability to offer an argument that supports one's conjecture is also a goal toward which mathematics students should aim. Conjecture without conviction is not likely to be any more valuable for students than proof without probable cause. Too often, proving has been introduced as a formal procedure that deals with statements formulated by others, but that the student may scarcely understand. If students have formulated their own conjectures out of patterns they have examined, they are more likely to argue for the conjecture. When students and their teachers are engaged in proving a textbook theorem, the object of the exercise is to come up with the "official" proof. When they engage in proving a student's conjecture, the object is quite different. It is to convince oneself and one's colleagues. Proofs should begin as plausible arguments. As students proceed through school, the standard of argument should be raised so that, eventually, reasoning is seen as resting on agreed-upon principles. All students need the experience of constructing arguments that will convince others, even if not all students are able to match the kind of official argument enshrined in textbooks.

Much of what we have been discussing in this chapter touches on matters of communication. In Chapter 2, we turn our attention to some issues associated with increasing communication in the mathematics classroom.

Notes

JUDITH D SEDWICK

2. Communicating Mathematical Ideas: Private and Public Discourse

The mathematics classroom could be a forum for the exchange of ideas about mathematics. But, all too often, students see only what ideas the teacher has and what the textbook authors have thought about. Rarely do they display their own thinking. In this chapter, we explore how we might get students' ideas and questions out into the open where they can become objects for reflection, refinement, discussion, and amendment. This "publication" of students' ideas occurs naturally through speaking and listening, but it can also unfold through writing and reading.

Mathematics deepens and develops through communication. Mathematical ideas become tangible when people find words and symbols to express them. The only way that mathematical culture can be passed on to new generations is through the medium of ordinary language, along with some special symbols. The interpretation of that language and those symbols requires learners to see them in much the same sense that teachers do. That demands sharing on the part of the teacher and the learner—a mutual construction of meaning. Building a mathematical classroom environment in which the mutual construction of meaning can be fostered is not a trivial matter, but it can reap substantial rewards for teachers and for students.

Notes

The traditional diet in mathematics classrooms consists of short answers to oral or written exercises. That kind of simple back and forth prevents a teacher from getting a deep sense of a student's understanding, and of how the student came to acquire that understanding. For that, a more sustained engagement between teacher and student is required. By providing opportunities for students to express themselves about the ideas at issue in a mathematics lesson—not just the next step or the last definition—a teacher can notice and counter confusions that students have accumulated. A teacher can also assist students to attain a more coherent understanding of what they are learning—one that will last beyond next week's test. Finally, by learning to express their ideas to one another, students can begin to appreciate the nuances of meaning that natural language often masks, but that the precise language of mathematical exposition attempts to distinguish.

Opening Windows: Teachers Learning from Students

When our grandparents were in school, recitation comprised a large part of their mathematics classes. Grandma would be called to the front of the class. She would be asked to put on the blackboard a problem she had previously solved, and then be asked to explain her solution to the rest of the class. Depending on how well it was done, her recitation might help her fellow students understand more about the problem and its solution. But, more important, whether or not it was done well, her recitation gave her teacher a window through which to view Grandma's thoughts, her reasoning, and her grasp of terms, or notations, or fundamental ideas. That view made it possible to diagnose and evaluate Grandma's understanding.

Today, formal recitations are absent from high school classes. Although some teachers ask students to put homework problems on the chalkboard or the overhead projector so that other students can see their work, students are seldom asked to put into words not only what they did but also how and why they did it. Few would advocate the return of the formal recitation, with the teacher—gradebook in hand—presiding over a by-the-book recital of "how you do it." However, there are moments and practices in teaching that allow teachers to glimpse beyond the homework paper or the short answer. Using these opportunities productively may take experimentation and practice, but the potential payoff for students and teachers is great.

In order to catch a glimpse of his students' thinking, Mr. Westerman often assigns a problem that can be solved by more than one method, and then has students present their various methods. In his algebra class, which was studying the solution of simultaneous equations, he assigned one of his favorite problems, which he knew would be accessible to students using different approaches.

In the barnyard, I have some chickens and some rabbits. I count 50 heads and 120 legs. How many of each type of animal is in the barnyard?

After an initial period of time, during which the class as a group discussed the problem to clarify interpretation, Mr. Westerman let the students work in pairs to solve the problem. As the students worked, Mr. Westerman circulated around the room, taking note of the different approaches being utilized, answering clarifying questions, and providing help to those who needed it. During his travels around the room, he asked José, Lora, Robert, and Karen to "publish" their solutions on pieces of white butcher paper that he provided. After reasonable time, Mr. Westerman invited the designated students to post their solutions on the blackboard and explain their approaches to the class. Mr. Westerman had carefully chosen the student publishers to represent a diverse set of methods, including some arithmetic, visual, and several algebraic approaches. In this way, his student presenters provided their classmates with an opportunity to see and discuss many different solutions to the same problem.

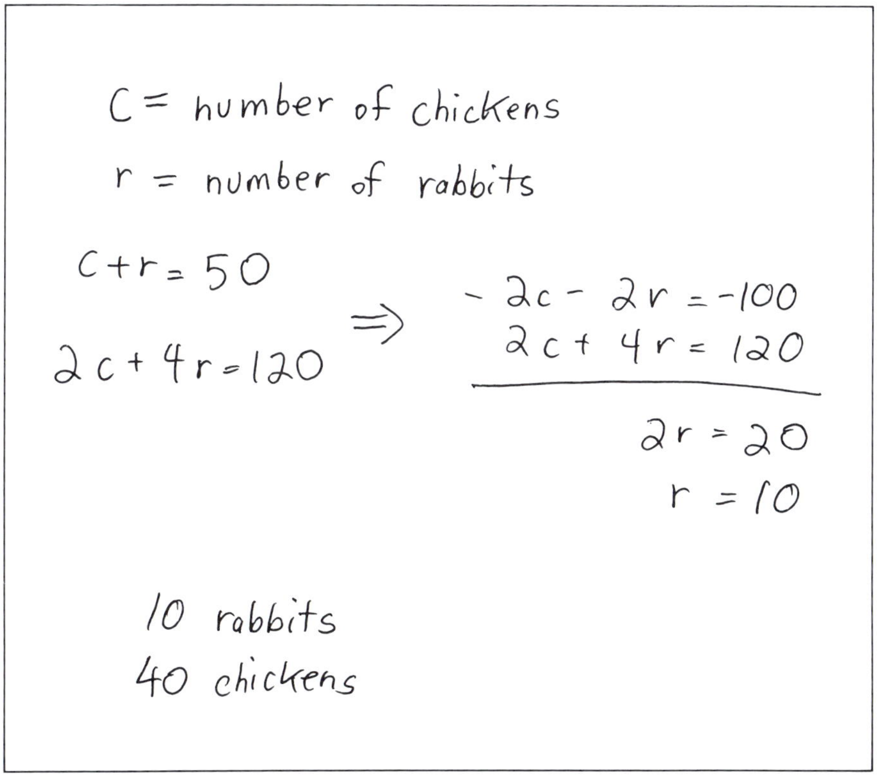

Notes

$$c = \text{number of chickens}$$
$$r = \text{number of rabbits}$$

$$c + r = 50$$
$$2c + 4r = 120$$

$$\Rightarrow \quad \begin{aligned} -2c - 2r &= -100 \\ 2c + 4r &= 120 \\ \hline 2r &= 20 \\ r &= 10 \end{aligned}$$

10 rabbits
40 chickens

Figure 2-1. José's Algebraic Approach to the Chicken-and-Rabbit Problem.

COMMUNICATING MATHEMATICAL IDEAS 17

Notes

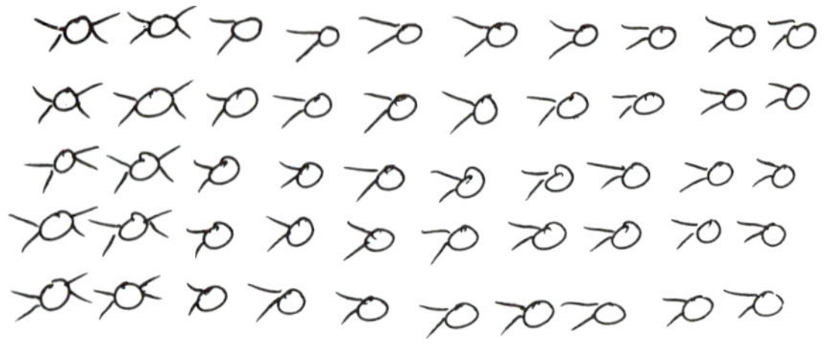

```
XX XX ⊅ ⊅ ⊅ ⊅ ⊅ ⊅ ⊅ ⊅
XX XX ⊅ ⊅ ⊅ ⊅ ⊅ ⊅ ⊅ ⊅
XX XX ⊅ ⊅ ⊅ ⊅ ⊅ ⊅ ⊅ ⊅
XX XX ⊅ ⊅ ⊅ ⊅ ⊅ ⊅ ⊅ ⊅
XX XX ⊅ ⊅ ⊅ ⊅ ⊅ ⊅ ⊅ ⊅
```

50 heads

Give everybody 2 legs, you use up 100 legs. Then you have 20 legs left over. Give them out in pairs of two (no one has 3 legs.)

You end up with 10 - 4 legged animals and 40 - 2 legged animals:

10 rabbits   40 chickens

Figure 2-2. Lora's Visual Approach to the Chicken-and-Rabbit Problem.

X = number of Chickens
$50 - X$ = number of Rabbits

$2X + 4(50 - X) = 120$
$2X + 200 - 4X = 120$
$-2X = -80$
$X = 40$

40 Chickens
10 Rabbits

Figure 2-3. Robert's Algebraic Approach to the Chicken-and-Rabbit Problem.

Chickens	Rabbits	# of legs
25	25	150
30	20	140
35	15	130
40	**10**	**120**

40 Chickens
10 Rabbits

Figure 2-4. Karen's Arithmetic Approach to the Chicken-and-Rabbit Problem.

Mr. Westerman first became interested in this technique when he read that this approach was used in many Asian schools. He thought it might make his Asian students feel more comfortable, and it might help his non-Asian students improve their ability to discuss mathematical ideas. He suspected the process would also provide some opportunity for him to learn more about his students' thinking and problem solving. He found that eavesdropping on students' conversations while they are working together in pairs or larger groups to solve a problem or explore a situation is a very useful way to learn about their thinking. But he was surprised to find that the process of publishing students' solutions also reaped benefits for him personally. Even though he was quite familiar with many of the problems he used—like the chickens-and-rabbits problem, the discussion often illuminated some facet of the problem that he had not seen or thought about previously, such as Lora's visual approach to a problem that he had previously thought of as being purely algebraic or arithmetic.

Not all of the exchanges that illuminate a student's thinking need to be public. When students are just beginning the study of a new mathematical topic, it is often assumed that students know very little about the topic. Actually, an inquiring teacher may find that, prior to their formal instruction, students often may have formed some rather complete, although possibly partially mistaken, ideas about the topic. One way to detect and

COMMUNICATING MATHEMATICAL IDEAS

Notes

examine the ideas students may have prior to formal instruction is for the teacher to provide a task, such as a writing assignment, that allows students to demonstrate what they know (or think they know) concerning a certain topic.

Mrs. Garcia was beginning a unit on circles in her tenth-grade geometry course. She knew that the students had studied circles in elementary school and had learned something about area, circumference, radius, and diameter in the seventh or eighth grade. She wasn't sure, however, what her tenth graders remembered, so she asked them, "Tell me everything you know about circles." She made this assignment during the last 10 minutes of class and collected the papers so she could look at them before class the next day. Figure 2-5 shows some examples of the responses she got.

Juan: "It's made up of a series of arcs that are all connected."
Monique: "A circle is a shape that has no points."
Louis: "The outside of a circle is called the circumference."
Ramona: "A circle is round."
William: "A circle measures 360°."
Eric: "Radius is half the distance of the diameter of a circle."
Cathy: "½ - 180°."
Rebecca: "$A = \pi r^2 (\pi = 3.1415927)$."
Michael: "It has 2 sides (inside and outside)."
Charles: "The center is the radius."

Figure 2-5. Responses to "Tell me everything you know about circles."

On the basis of her students' responses, Mrs. Garcia knew that there was a wide range of knowledge about circles in her class. Some of her students had remembered a good deal about circles (e.g., radius is half the length of the diameter, area is πr^2), others gave no evidence of recalling any important mathematical characteristics of circles, and still others had some errors (e.g., the center is the radius) or potential misconceptions in their knowledge of circles (e.g., a circle is a shape that has no points).

After reviewing the responses, Mrs. Garcia decided to use the students' responses as a jumping-off point. Keeping her plans for the unit on circles firmly in mind, she chose 10 different responses (the ones shown in Figure 2-5) and typed them on a ditto sheet. She left space after each statement and challenged the students to decide whether or not they agreed with the statement, and to provide a justification for their agreement or disagreement. While most of the class worked on this assignment, Mrs. Garcia worked separately with a small group of students—those whose responses revealed most of the misleading or erroneous information. Because of their confusions about circles and their high risk for difficulty in the next unit,

Mrs. Garcia met with these students and provided special help before they floundered in the circles unit. Thus, the initial, brief writing assignment allowed Mrs. Garcia to sample her students' prior knowledge about circles, and it also helped her to diagnose a few misunderstandings, identify students who might need special attention, and provide an interesting way for the entire class to review their prior knowledge of circles.

Mrs. Garcia gave the assignment again at the end of the unit and used the students' responses to assess informally how their concept of circles had changed during the unit of instruction. Adding this information to the more formal information from the unit test helped her to assign students' grades, and also to evaluate the effectiveness of her instruction and plan how she would revise her teaching of this topic the next year.

When students are asked to speak or write about their ideas in a mathematics class, a teacher has an opportunity to become a learner. Although this process often unearths troubling evidence of students' confusions or misunderstandings, for which there may be no immediate remedy, a teacher at least becomes aware, at a deeper level, of the common understandings or misunderstandings that are characteristic of her students. Armed with this new-found knowledge, teachers can use this information to guide their instruction and their own reflections about its effectiveness.

Getting It Straight: Students Learning from Themselves

Our thoughts are ephemeral creatures that won't be pinned down until we articulate them in speech or writing. It is only when we have said or written them, and then have reflected on them, that we know what we are thinking. In far too many mathematics classes, ideas pass from the textbook page or the teacher's lips, through the students' eyes or ears, and into a black hole. They remain in the students' minds only long enough to permit the solution of some immediate exercise or problem. Come back another day to ask about the ideas, and they have vanished. We cannot retain much of the mathematics we have seen or heard if we have not appropriated it as our own—if we have not thought about it in some fashion. Reflection is a way to grasp our thoughts, and it is enhanced if we can get it out into the open or pin it onto the page. Once it is there, we can examine, repair, or augment our thinking.

At the end of class, Mr. Johnson often takes five minutes to ask his students to write what they have learned that day and what they still have some questions about. By glancing over the papers, he can make a quick appraisal of students' confusion and points of clarity. He has found that his students like this practice. It puts them in charge of clarifying their own thinking and diagnosing their own misunderstanding.

Mr. Johnson also learns a great deal from his quick review of what his students have written. This helps him to plan the quick recap that will be at the beginning of tomorrow's lesson, to select or create exercises for

Notes

Notes

subsequent homework assignments, to pose useful questions for the end-of-chapter review, and to identify students who may need special help. But he feels that the greatest benefit of this technique is that it places much of the responsibility for learning on the shoulders of his students, rather than solely on himself. By asking his students what they have learned in today's lesson, he reminds them that they, too, are responsible for the results of each lesson.

One way of giving students a more substantial perspective on their intellectual progress is to ask them to keep a journal of their thoughts about the mathematics they are learning. This "learning log" becomes a place where students can express their puzzlements, frustrations, and triumphs. They can write about their joy at finding a solution, or their anguish at finding it incorrect. Each day, they can set down a portion of their intellectual journey. By looking back over the entries, they can see how far they have come this month or this year. Some teachers ask students to use the log as a kind of mathematical diary in which they can record their daily thoughts. The log then functions much as Mr. Johnson intended—revealing what they have learned today—but it also serves as a record of learning over time. Other teachers ask students to record their thoughts and feelings as they solve homework problems or work on extended investigations. Students might be encouraged to write in the log their solutions to special problems, along with annotations concerning their strategies and evaluations. When students are asked to write up their solutions to a problem, they can be encouraged to comment on their false starts, hesitations, and uncertainties. They are, after all, communicating with other human beings who are interested in their work. Of course, the learning logs, like most diaries, should have limited circulation. Some teachers choose to read the logs only at mid-year and end-of-year points; others read them more frequently. Some teachers never read the logs at all, preferring instead to have the students keep the journals only for their own use. No matter how often the journals are read, their use can promote a more reflective attitude often described as increased *metacognitive* awareness on the part of students. Teachers of students for whom English is not their first language may find that logs can help those students to formulate their ideas, and the expression of those ideas, before displaying them in class, or at risk on a test.

A Community of Learners: Students Learning from Each Other

A popular image of the mathematician is of someone isolated in a paper-strewn study. But mathematicians are also social beings. Mathematical ideas gain their validity by being accepted in the community. The common view in most mathematics classrooms is that a proof is a proof. The emergent view in mathematics, however, is that a proof is not a proof until it has been accepted by the community of mathematicians (Tymoczko,

1986). The controversy and disagreement about the acceptability of the computer-based solution of the famous, and long-unsolved, "Four Color Problem" (Appel and Haken, 1977) provides a contemporary illustration of this phenomenon within the mathematical community (Peterson, 1988).

Even when we are not trying to produce publishable solutions to important mathematical problems, we need to test our mathematical ideas against those of others so that we can be confident that they are correct. The only way we have of knowing what our thinking is like is by comparing it with that of others. The major pedagogical implication of this view is that mathematics classrooms should be communities of learners. Within communities, the need for communication is obvious. Within mathematical communities, communication in the form of discussion, argument, proof, and justification is natural.

It is well documented that students have great difficulty learning to produce mathematical proofs (Senk, 1985; Senk and Usiskin, 1983). Some part, but not necessarily all, of that difficulty is due to the logical requirements of mathematical proof. Another factor that makes proof difficult for many students is their lack of a sense of why one would ever engage in proving something. Isn't it enough to "see" the truth of a statement? If the teacher has given a statement to be proved, then it must be true. Why bother? Isn't proof really much ado about something obvious? Proof as a private exercise in sheer logical reasoning apparently does not strike many secondary school students as vital, but proof as an exercise in justification and validation in a community of learners holds considerably more appeal.

When students are challenged to think and reason about mathematics, and to communicate the results of their thinking to others in the form of a justification or proof, they are faced with the need to state their ideas clearly and convincingly. The act of communicating ideas within the culture of mathematics creates both the need for and the value of mathematical proofs. A proof that emerges as a student's attempt to convince skeptical peers of the power of a self-generated assertion has an importance that the demonstration of an "obvious" theorem from a textbook rarely has.

Much is learned by constructing a proof or a justification argument. Occasionally, an assertion turns out to be untrue as stated, and the assertion has to be modified. Such an experience demonstrates vividly the rough, tentative nature of mathematics in the making, as opposed to the polished shine of finished mathematics. Proving an assertion can also lead to insights into further statements that also appear to be true. And the presentation of a proof to the community often leads to further refinement and modification, even correction, in order to improve the clarity and precision of the justification.

Working in pairs or small groups also provides students with an opportunity to validate their reasoning and their conjectures. They can work through their disagreements to a consensus about what works and what makes sense. They can hear their poorly formulated expressions phrased more precisely by their peers. They can help their colleagues grapple with ideas they themselves have just begun to understand. Many high school

Notes

Notes

mathematics teachers are skeptical about the benefits of group work. Cooperative classroom groups are viewed as difficult to organize and manage: group work is noisy; it takes time; it often seems to be leading nowhere. Many teachers believe that it is much simpler just to tell students, rather than to leave them floundering. Yet for those teachers who are willing to work through the problems of managing groups, the benefits in student understanding can be substantial and lasting.

Of course, students need not be in small groups to have a dialogue with one another. In the hands of a sensitive teacher, a student presentation to the whole class of a problem just solved, or papers outlining a proposed proof that are exchanged to be read and given written commentary, can be vehicles for students to reach new levels of understanding. The key lies in the spirit of shared inquiry that the teacher is able to foster and sustain. Certainly, the process of publishing solutions discussed above in the context of Mr. Westerman's class is a good example of a classroom interaction structure that allows students to learn a great deal from each other during a process of semi-formal presentations to an entire class.

3. Transforming Ordinary Situations Into Extraordinary Opportunities

The picture of a richer, more vigorous school mathematics that has been painted in the earlier chapters does not require a new curriculum, new textbooks or teaching materials, or different students. Much can be done with what we have now. What we need is to look for appropriate opportunities for thinking and communication in the material we already teach. In this section, we offer some examples from teachers who have made such opportunities by transforming ordinary problems and situations into extraordinary occasions for developing and promoting mathematical thinking.

Opportunities for Mathematical Speculation

The popular image of mathematics emphasizes its precision and finely tuned final products. In this view, there is little room for such activities as wondering and speculating. However, such open-ended thinking is actually quite characteristic of mathematics. For example, important generalizations can often be generated through the process of speculating about how

Notes

a given result might be extended in new directions. We need to provide students with some opportunities to engage in these more open-ended forms of mathematical thinking.

Modifying common textbook problems to make them more open-ended can be a plausible entry point for introducing speculation. For example, Mr. Ruiz frequently changes the problems presented in his students' geometry textbook to more open-ended and challenging versions. Some examples of the textbook problems and his revisions are shown in Figure 3-1. Mr. Ruiz has found that when his students work on the open-ended versions they think about the problem longer, they pose more problems and find more solutions, and they make more interesting and important connections among mathematical ideas. Mr. Ruiz sometimes uses an open-ended problem as an exercise to help students review for a unit test, but he has also used an open-ended question (like Problem 1B in Figure 3-1) as a way of reviewing what students already know before a new unit is started and as a way of solidifying learning during an instructional unit.

Ms. Smith also likes to pose open-ended problems for students in her geometry class to think about. She relies on examples that she has collected from articles in professional journals and from workshops she has attended.

	Goal-Specific	Non-Goal-Specific
1A	Two parallel lines are cut by a transversal, as shown in the figure below. If angle 2 is 30°, find the measure of angle 1.	1B Two parallel lines are cut by a transversal, as shown in the figure below. If angle 2 is 30°, find the measure of as many of the other angles as you can.
2A	The radius of a circle inscribed in a square is 6 inches, as shown in the figure below. Find the area of the square.	2B The radius of a circle inscribed in a square is 6 inches, as shown in the figure below. Find out all you can about the circle and square.

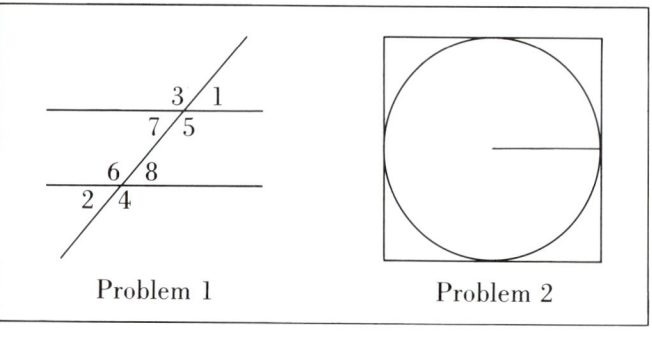

Figure 3-1. Examples of Open-Ended Problems Used by Mr. Ruiz.

One problem that she has used frequently with her geometry classes involves identifying the shapes that can be made by a straight line passing across a square. It is an engaging problem because it is accessible to all her students, yet it leads to interesting discussions and speculations that deal with important geometric ideas. The following excerpt makes this clear:

> *Smith:* Imagine a line passing across a square. What shapes could be formed? The line can pass across the square in any direction we want. I'll draw a square here on the board, and I need someone to come up and draw a line passing across the square. Jim, please come to the board and draw a line across the square.
>
> (Jim draws a line as directed.)

Smith: OK. What shapes are formed?
Jim: A trapezoid.
Smith: Only one?
Jim: No, two.
Smith: OK. Everyone should use the sheet of paper I gave you to find as many different possibilities for shapes made by a line across a square.

> (Students work individually for about five minutes on a sheet filled with squares.)

Smith: Everyone team up with a partner and compare the shapes you have found. Compile a master list of all the different possibilities you have identified. While you're working, I'll pass around an overhead transparency, and each team should add one *new* answer, if you have one, to the set that is already there.

> (Students work in pairs for about 10 minutes, comparing their work and compiling their list of answers. As the transparency passes around the room, students carefully review the set of answers and add new ones whenever possible.)

Smith: OK, let's look at the answers we have come up with.

> (Smith projects the transparency. She and the students identify and discuss each answer to be certain that it is valid and new.)

Smith: Are there any more answers that we should add to our list? Melanie?
Melanie: I'm not sure this is new, but what about the square itself? Could that be an answer?
Smith: What do you think? Can anyone see some way that the square itself could be an answer? Mark?
Mark: Sure, if you have the line be like a side of the square. Is that legal?
Smith: What do you think? Does anyone have an idea about this? Did any of you include the square on your list of possibilities? Sam?

Notes

Notes

Sam: I did, because I thought it was, like, a special case or something. You know, when the line just happens to line up with the side. You didn't say the line couldn't be that way, so it seemed OK.

Smith: That's good, Sam. The directions did not disallow it, and so we could consider this a special case. So that was a wonderful suggestion, Melanie, and we need to add it to our list. Anything else? Rebecca?

Rebecca: I'm not really sure about this, but when we make triangles with the line, can't we make some special triangles sometimes? I mean, is "triangle" really saying enough?

Smith: What do you mean by "special" triangles?

Rebecca: Like equilateral, or isosceles, or even 30-60-90.

Smith: Oh, I see. That's a good question. I want each of the pairs to go back and look at your answers to see if you can find examples of any special triangles that are formed. Think about how you could convince yourself that the triangle is really special.

(Students work in pairs examining their lists to see if they can find any "special" triangles.)

Smith: OK. Who has some ideas to share with us? Jenny?

Jenny: Well, we decided that you can't ever have an equilateral triangle, because there's always a right angle in there somewhere. And we had that problem before, when you asked us if an equilateral triangle could have a right angle, and we found out that it can't.

Smith: Is that clear to everyone? Jenny and Deb have decided that, in any triangle formed in the square by the line passing across it in this way, one of the angles has to be a right angle. Since you can't have a right angle in an equilateral triangle, then there can be no equilateral triangle in our set of triangles. Any questions? Mike?

Mike: It's easy to get an isosceles triangle—just put the line across as a diagonal.

Rebecca: Yeah, and that is really an isosceles *right* triangle. We got that one.

Smith: What are the angle measures in an isosceles right triangle?

Rebecca: OK, so another way to call this triangle is a 45-45-90 triangle.

Smith: Good. Anything else?

Sam: If you put the line so it hits a corner and the midpoint of a side, then you get a special triangle—a 30-60-90 triangle.

Smith: Is that clear to everyone? Sam, please draw that one up here on the board for us and explain your reasoning about the 30-60-90 triangle.

(Sam draws a line connecting a vertex to the midpoint of a nonadjacent side.)

Sam: Well, if the side of the square is one, then this piece is one-half of that, and this is a right angle, and so you get 30-60-90 for the angles.

Smith: Any questions or comments? Monica?
Monica: I thought the hypotenuse was supposed to be twice as long as a leg to have a 30-60-90 triangle. Here we have one leg twice as long as the other leg. I think we saw that this wasn't right back when we studied 30-60-90 triangles.
Sam: She's right. I blew it! This sure seems like a special triangle, but it's not 30-60-90.
Smith: Did anyone have a 30-60-90 triangle in their collection of possibilities? (No response.) Do you think it's possible to get a 30-60-90 triangle with a line passing across a square? Why don't you work on that problem for homework. Also, I want you to think about another special issue—congruence. When are the shapes formed by the line across the square congruent? Finally, I would like you to think about how to generalize this problem. We have tried to generalize some of our other problems, and I want you to think about what some appropriate generalizations might be for this problem.

This excerpt from Ms. Smith's class illustrates the kind of conversation that can take place in which students conjecture about and propose mathematical possibilities, and evaluate their arguments and assertions. Ms. Smith started the conversation at a level of visualization that was comfortable for all her students, but she continually prompted a discussion of more sophisticated mathematics and mathematical issues, and wove it all together in the conversation. The pace was not too rushed—she gave students time and space to think, to share their ideas, and to contribute their best efforts (both individually and collaboratively) to an increasingly complete resolution of this open-ended problem. She treated tentative suggestions and conjectures seriously by encouraging the class to examine and probe them. Even errors were interesting to Mrs. Smith; she treated them as reasonable speculations and explored them for their learning value. The assigned homework grew naturally out of the unresolved issues related to the problem, and contained in it the seeds for still later generalizations and speculations.

Ms. Smith's final homework task—asking her students to generalize the problem—would surely please her colleague Mr. Durant, who teaches at another high school in the same district. A few years ago, he read a book called *The Art of Problem Posing* (Brown and Walter, 1983) and began to incorporate problem posing and generalization activities into all of his mathematics classes. One of his favorite activities involves the Pythagorean Relation: The sum of the areas of the squares on the legs of a right triangle is equal to the area of the square on the hypotenuse. He enjoys asking his intermediate algebra students to generalize the Pythagorean Relation, which they have already studied in elementary algebra and in geometry. Over the years, he has been amazed at the variety of generalizations that the students can produce—including several different generalizations in three dimensions, and many generalizations related to constructing other figures (e.g., semicircles or regular polygons other than squares) on the sides of a right triangle.

Notes

Notes

Mr. Durant remembered the first time he asked his students to do this activity, because a student generated a conjecture about which he was unsure.

> I remember Mark, one of the best students in the class, saying that he thought [the Pythagorean Relation] could be generalized to any regular polygon. I had seen a proof for equilateral triangles and regular hexagons, but I didn't know if it was true for every regular polygon. And I sure didn't know how to prove or disprove it. I felt pretty uncomfortable not knowing the answer. After examining proofs for the cases of the equilateral triangle and regular hexagon with the entire class, I had Mark state his conjecture. Since Mark was the best student in the class, and since we had seen a few specific cases, everyone was prepared to accept it. But I warned them about the need to develop a convincing argument, and made it available as special extra-credit project. Mark and two other students got "hooked" on the project, and they were finally able to prove the general result by finding and using a formula for area using the apothem—a formula that was not in the curriculum I was teaching.

Mr. Durant's class had learned at least two important lessons: the truth of a mathematical assertion is not always known in advance, and, with persistence and reflection, students can often engage in significant mathematical thinking that goes well beyond what they are normally asked to do. Mr. Durant, too, had learned an important lesson.

> After that experience, I decided that it was not such a serious problem if I didn't always know whether or not my students' conjectures were valid, so I just routinely ask them to justify their own thinking. This approach turned out to be helpful for some of the three-dimensional generalizations. A few years ago, one student turned the generalization of the Pythagorean Relation in three dimensions into a winning project at the district science fair. In fact, I like this approach so much that I find myself continually acting skeptical and challenging my students to justify their mathematical assertions.

Opportunities for Thoughtful Mathematical Analysis

Mathematical analyses of interesting situations can often arise as a result of considering real-world applications, such as the problem of planting replacement trees discussed earlier. However, many thoughtful analyses can also arise naturally from problems already present in the secondary school curriculum. With exploration and expansion, even ordinary textbook problems can be used as springboards for careful analysis.

Mrs. Herrera noted that her students had some difficulty in setting up equations to solve "age" problems in elementary algebra. To get them to focus on the relationships she thought were at the heart of her students' difficulty, Mrs. Herrera asked her students to think about how the relationship between two persons' ages changes over time. In particular, she asked them to think about how the difference between their ages changes and how the ratio of their ages changes over time. This turned out to be very difficult for her students: what they could observe and the questions they could generate were disappointing. Based on what she perceived to be her students' frustrations with the abstractness of the task, Mrs. Herrera decided to try grounding the ideas in a more concrete problem setting.

The next day, Mrs. Herrera gave her students the following problem from their elementary algebra textbook: "Pop Fligh is 8 times as old as his son, Hy. In 2 years, Pop will be only 6 times as old as Hy Fligh. What are the ages of Pop and Hy now?" She set everyone to working in pairs to (a) solve the problem and (b) complete a chart by listing the ages of Pop and Hy for five years before and five years after today. Although they had a little difficulty solving the problem, most of the students, working collaboratively, were able to find a solution for the problem and complete the chart. Then Mrs. Herrera asked half the class to fill in the "Diff" row in the chart by calculating the difference between Hy's and Pop's ages for each year. The other half of the class was asked to complete the "Ratio" row in the chart by calculating the ratio of Pop's age to Hy's age (Pop:Hy) for each year.

	Before					Now		After			
Year	−5	−4	−3	−2	−1	0	+1	+2	+3	+4	+5
Pop	35	36	37	38	39	40	41	42	43	44	45
Hy	0	1	2	3	4	5	6	7	8	9	10
Diff	35	35	35	35	35	35	35	35	35	35	35
Ratio	??	36	18.5	12.7	9.8	8	6.8	6	5.4	4.9	4.5

After the problem was solved and the chart completed, Mrs. Herrera led a brief discussion about the answers to the two different problems. No one was particularly surprised to find the constant difference between the ages, but many expressed considerable surprise about the ratio data. In particular, they noticed that the ratio was not constant, that it was undefined for the first year in the chart, that it changed very dramatically over the first few years and then more slowly, and that it decreased in value each time. There was also some interesting discussion about the use of a single decimal number to express the ratio. By the end of the discussion, even those students who originally thought that "you have to have two

Notes

numbers in a ratio" agreed that the use of a single number simplified the analysis of the relationship.

Mrs. Herrera decided to push further. She asked her students to make a graph by plotting (year, difference) ordered pairs. She also asked them to predict and sketch what the graph would look like before finally plotting the points. Most of her students had little trouble predicting the "flatness" of the graph because of the constant difference, although some students predicted a vertical rather than horizontal line.

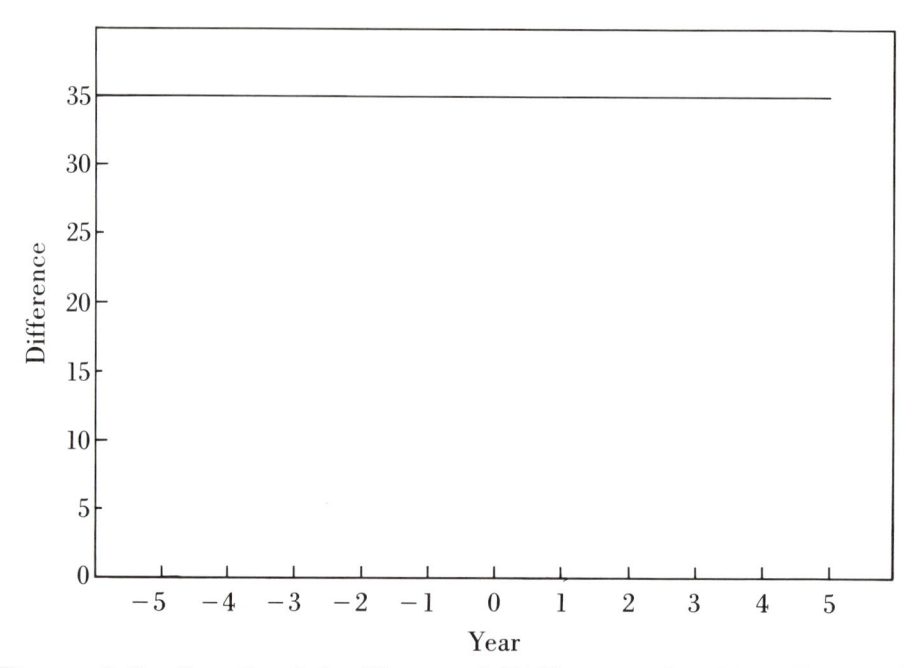

Figure 3-2. Graph of the Year and Difference for Mrs. Herrera's Age Problem.

Following a discussion of the relationship between the graph, the chart, and the actual ages of the people, Mrs. Herrera felt her students were ready for a more challenging problem. For homework, she asked the students to make a graph by plotting (year, ratio) ordered pairs. As with the task solved in class, she also asked them to sketch their prediction before actually plotting the points.

When the students came to class the next day, they were more than ready to discuss their graphs. There was genuine excitement: Not only wasn't the graph constant, it wasn't a straight line at all!

There was a lively debate about how the graph should be extended, so Mrs. Herrera assigned as homework the task of making a prediction, extending the data in the chart for another 10 years into the future, and plotting the points that extended the graph. The following day, the class discussed the behavior of the curve representing the ratio of the ages. They noted that the values in the chart and the graph of the curve settled down close to, but never reached, the ratio of 1. They could see why that

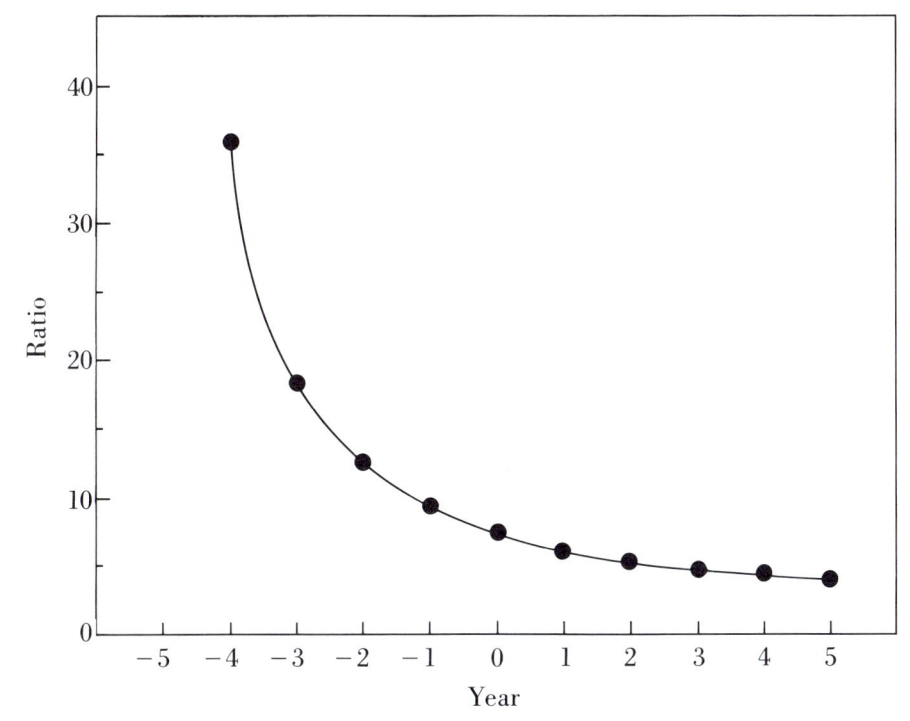

Figure 3-3. Graph of the Year and Ratio for Mrs. Herrera's Age Problem.

made sense, as one student remarked, "It [the ratio] could never be 1 because there is always that difference in their ages. We already saw that it never changes."

Although her class had invested considerable time in this set of problems, Mrs. Herrera was more than satisfied with the results. She had begun with the simple hope that this experience would help her students solve "age" problems more successfully, but she was delighted that there were other very important mathematical benefits from the experience. What had begun as an attempt to help students solve ordinary textbook problems had turned into a startling mathematical investigation. For several days, her students had been engaged in analyzing and discussing various interesting, important, and fairly sophisticated mathematical ideas. Their exploration had introduced them to a constant linear function, to the representation of a ratio as a single number, to a monotonically decreasing nonlinear function, and to an intuitive notion of an asymptote. Mrs. Herrera was certain that she would be able to refer to this experience in subsequent lessons to build on some of these ideas.

Mrs. Herrera was excited and pleased enough to discuss her experience with some of her colleagues. She was subsequently invited to give a talk at the local mathematics conference. One of the other sessions at the conference was given by Mr. Jones, a teacher from another high school in the area. His talk focused on ways of transforming geometry problems to make them more interesting and challenging for students. At the confer-

Notes

ence, he discussed his experience with several classes he had recently taught, ranging from tenth-grade geometry to precalculus.

I was sent a problem by a colleague I met during a trip to Japan. He said that he had used this problem with his Japanese secondary school students and that they had produced some interesting solutions, so I decided to try it with my precalculus students.

Imagine that you are given a piece of paper with an angle drawn on it. (See Figure 3-4A below.) *Your teacher asks you to take the paper home and construct the bisector of the angle. On the way home on the bus, the paper is ripped, and the vertex of the angle is torn off.* (See Figure 3-4B below.) *How can you complete the assignment using only the torn piece of paper you have left?*

After explaining that it was not "legal" to put the "torn" angle on another piece of paper and extend the sides to find the vertex again, I assigned this problem as an extra-credit project investigation to be done by individuals or teams of students. I told them that the problem had come from a colleague in Japan, and they seemed highly motivated to do their best to solve the problem. They had been reading newspaper stories about how poorly United States students do on mathematics tests compared with Japanese students, and they wanted to show that they could do something with this problem.

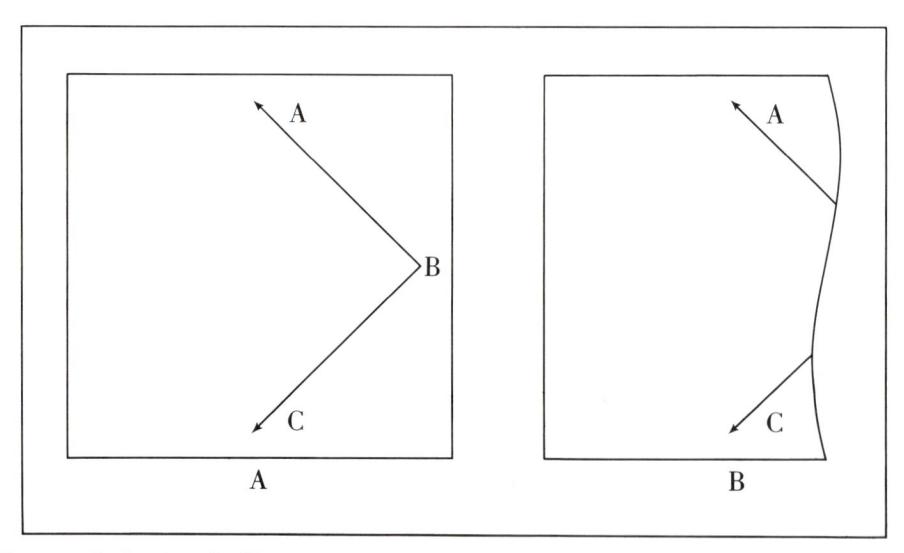

Figure 3-4. Angle-Bisector Problem Used by Mr. Jones.

Mr. Jones went on to explain how he often assigned longer term projects like this one to give students a chance to investigate some mathematical problem or ideas, and to earn extra credit. After a few years of doing this in different ways, he had learned that encouraging the students to work in pairs increased the probability that some good ideas or solutions would

34   THINKING THROUGH MATHEMATICS

emerge. The collaborative approach seemed to fit the way people really work on problems outside school.

I was quite pleased with the variety of solutions proposed by the students, and the discussion of solutions provided useful review of many geometric ideas. The students used a great deal of their geometric knowledge and really were quite clever in their approaches. (See Figures 3-5 through 3-7 for examples.) Kerwin and John, whose solution was judged by the class to be the easiest to understand, said that they got the idea for their solution by trying to be very simple-minded about angle bisector—by trying to think of its basic meaning as a set of points halfway between the sides of the angle. They used a clever approach to find two such points and knew they could then draw the line that would represent them all. After discussion, the class realized that all the solutions really used this basic meaning of the angle bisector in some fundamental way. Tracie's solution was based on the same idea, but she used knowledge about parallel lines

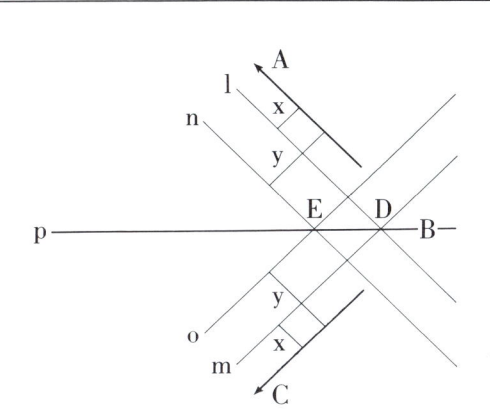

1. Construct line l parallel to AB at distance x from AB.
2. Construct a line m parallel to CB at distance x from CB.

The point of intersection of these lines D, is one point on the angle bisector.

3. Construct line n parallel to AB at distance y from AB.
4. Construct line o parallel to CB at distance y from CB.

The point of intersection of these lines E, is another point on the angle bisector.

5. Draw line p through points D and E. Line p is the angle bisector.

Figure 3-5. Kerwin's and John's Solution to the Angle-Bisector Problem.

Notes

in a somewhat different way—a way that the class found more difficult to understand. The solution produced by Ramona and Melanie was nice because they made use of the incenter—the point of intersection of the angle bisectors—and an inscribed circle. Their solution was viewed by the class as the most sophisticated one.

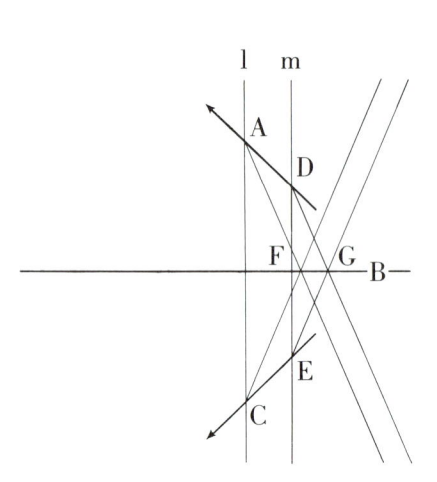

---

1. Draw transversal l through BA and BC.
2. Draw transversal m through BA and BC parallel to transversal l.
3. Bisect angle A and angle C, the base angles of triangle ABC.
4. Point F, the intersection of the angle bisectors, must be one point on the bisector of angle B.
5. Bisect angle D and angle E, the base angles of triangle DBE.
6. Point G, the intersection of the angle bisectors, must be one point on the bisector of angle B.
7. Draw a line through points F and G. This line is the bisector of angle B.

Figure 3-6. Tracie's Solution to the Angle-Bisector Problem.

Mr. Jones said that he thought the problem provided many students with an opportunity to explore an interesting problem, and also to review a fair amount of geometry at the same time. But it was more than a simple review. Several students remarked to him that, although they thought they already knew the topic very well, this problem pushed them to think more about what an angle bisector really was. This task—a slight variation on the usual assignment to construct an angle bisector—allowed students to engage in a thoughtful analysis of the basic meanings of angles, angle bisectors, loci of equidistant points, and other fundamental geometric ideas.

In fact, the problem worked so well that Mr. Jones decided to challenge his tenth-grade geometry students with the problem, near the end of their year-long course, in order to stretch their understanding of geometric

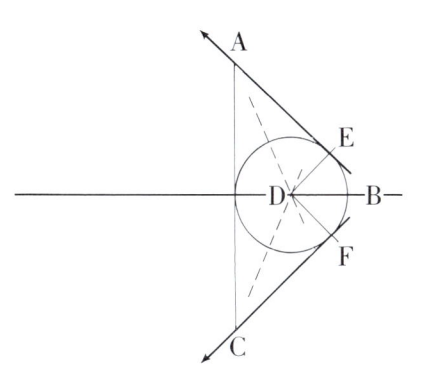

1. Connect points A and C.
2. Construct the bisector of angle A.
3. Construct the bisector of angle C.

Point D, the point of intersection of the angle bisectors, is the center of the inscribed circle.

4. Construct a line passing through point D and perpendicular to AB.
5. Construct a line passing through point D and perpendicular to CB.
6. Draw radius DE and DF.
7. Bisect the central angle EDF.
8. The bisector of central angle EDF is also the bisector of angle B since we have constructed congruent triangles DEB and DFB.

Figure 3-7. Ramona's and Melanie's Solution to the Angle-Bisector Problem.

constructions. Unfortunately, the problem appeared to be too difficult for the students in his class.

> No one found a complete solution, and a few students were really frustrated and annoyed with the problem. I even got one call from an angry parent who was unable to solve the problem and was frustrated. A few students did come up with the idea of folding one side of the angle onto the other side in order to locate the bisector on the crease line. One student even mentioned that this reminded him of something he had done in junior high school with a red plastic mirror. Although their intuitive idea of a bisector was just fine, I was basically quite disappointed that no one was able to find a really rigorous construction.

Mr. Jones then decided that he would be satisfied if the students could justify and analyze the solutions, even if they couldn't independently produce them.

Notes

I gave the students a packet of solutions, including the folding solution from this class and several other solutions produced by the precalculus class, and I grouped them to work in pairs, and challenged them to justify each of the solutions or explain why they were incorrect. We started this during the last half of one class, then they worked on it independently at home, and finally together again for the first half of the next class. And this worked much better, with the burden of solution lifted off their shoulders. Some of the students really surprised me with how well they were able to reason about the various solutions.

Most of the students were able to produce a good explanation or justification for the intuitive folding solution, and that helped them to see the angle bisector as locus of points. Some students were able to develop solid justifications for the other solutions, although many students found this difficult. The class discussion of the different solutions and attempted justifications not only gave Mr. Jones an opportunity to discuss the nature of geometric construction and to review a fair amount of fundamental geometric knowledge, but it also provided a forum in which to discuss and analyze what is involved in making a mathematical argument.

Opportunities for Convincing Oneself and Others

Mr. Jones provided the students in his class with an opportunity to see, hear, debate, and evaluate mathematical explanations and justifications. In other words, Mr. Jones's classroom became a place in which the emphasis was less on memorizing and producing answers and more on analyzing and becoming convinced.

Many of the problems considered in this or earlier chapters also provide occasions for students to engage in the process of convincing themselves or others of the validity of their mathematical ideas or assertions. For example, Mr. Jones clearly gave his students an opportunity to examine seriously how one convinces oneself and others of the correctness or reasonableness of a solution or approach. His classroom was transformed into an arena in which convincing and justifying became the central focus of attention, rather than a peripheral matter. Moreover, we have seen how Ms. Smith's "line across the square" problem led her students not only to conjecture about possibilities but also to verify and validate their conclusions about the possible shapes. In the class discussion, we saw Sam revise his assertion about the 30-60-90 right triangle on the basis of evidence provided by a classmate. And in the discussion of Mrs. Herrera's "age" problem, we saw her students predict the behavior of a graphed curve and justify their predictions on the basis of the data. Moreover, they were able to revise their individual and collective thinking about the relationships implicit in a problem situation.

It is important for students to realize that revision of their thinking is legitimate and appropriate. Since we very often present only polished, final arguments in mathematics classrooms, students have too few opportunities to see the importance of revision as a necessary component of reasoning. The philosopher Imre Lakatos also had such concerns about formal presentations of mathematics: "Naive conjecture and counterexamples do not appear in the fully fledged deductive structure: the zig-zag of discovery cannot be discerned in the end product" (1976, p. 42).

Although our students are taught by teachers of other subjects to develop "drafts" during the process of producing compositions and essays, we mathematics teachers rarely encourage the same behavior when they are constructing mathematical arguments. In fact, we spend so much time in our classes developing the "punctuation and grammar" of mathematics that we too rarely give our students the chance to do any "composing." When the focus shifts to allow more composition work in mathematics, then it becomes more natural to think of revision as a natural part of the process. Moreover, when more emphasis is placed on convincing rather than on memorizing, it also becomes natural to think of revision as a necessary component of the process. Convincing oneself or others is often a slow, repetitive process. We need to remember that—both for our students, as we plan and construct classroom activities, and for ourselves, as we set out to convince our students that mathematics is a rich, interesting subject that deserves their prolonged intellectual attention, not just their first-draft thinking.

Notes

4. Getting Started

The discussion in the earlier sections of this book has emphasized the importance of inquiry and communication in the secondary school mathematics classroom. Although most teachers would acknowledge the value of inquiry and communication, many have difficulty imagining how to make these things actually happen. Standard textbooks do little to suggest how more communication between teacher and students or among students might be fostered, nor do most textbooks provide much guidance about stimulating inquiry. In this chapter, we turn our attention to some very specific practical suggestions that may help teachers open up lines of inquiry and communication in their mathematics classrooms.

Providing Opportunities: Opening Up Lines of Inquiry and Communication

Many teachers believe that changing the classroom organization to allow more time for students to work in cooperative groups affords the best opportunity for stimulating both inquiry and communication. When students are working in groups, the teacher is free to move about the room and engage in dialogues that provoke or probe. Moreover, group work also

Notes

affords students the opportunity to think out loud about mathematics with their peers. Many mathematics teachers, however, do not feel comfortable with the notion of having students working in groups. For those teachers, communication opportunities can often be found through large-group discussions in class, and individual conversations with students while they are engaged in seat work. Regardless of the form of classroom organization employed by the teacher, when students work on an open-ended problem or an extended investigation, there must be opportunities for communication between teacher and students, either during class or during a free period, at lunch hour, or before or after school.

Timely Advice

The journey toward creating a mathematics classroom atmosphere that fosters communication and inquiry is sometimes a long one. Although the destination is clearly worth the trip, the path one travels to reach the goal is not always smooth. As is true of many of life's journeys, the trip will be more enjoyable if you take someone along with you. Most teachers find that innovation is easier if they are not alone in their efforts. Not only can a colleague provide much needed moral support during difficult times, but the colleague can also serve as a knowledgeable "sounding board" for ideas before they are tried in the classroom. "Such interchange provides intellectual refreshment and places teachers in the roles of partners in the process of education" (National Council of Teachers of Mathematics, 1989b).

On Changing the Rules: Expect Some Resistance

By the time students have reached the secondary school mathematics classroom, they have had substantial experience in mathematics classes—an experience that has built expectations about how mathematics classes should be conducted. It is not likely that this set of expectations includes the kind of communication and inquiry discussed in this book. More likely, students expect each class to consist of the teacher leading a review of the previous night's homework, then giving a short lecture on today's topic, and working a few examples at the board; after that, they expect to be assigned a few problems for seat work, and then given some time to begin their homework. Although students may not be particularly fond of this kind of mathematics classroom, they are comfortable with it and generally expect to meet it time and time again. The notion of having an extended discussion about an interesting problem, considering multiple solutions to the same problem, working in small groups and recording ideas, or writing about mathematics will seem foreign to most students.

Many students have reached an implicit understanding with their mathematics teachers: if the teacher will simply tell them the rules and when to use them, the students will follow the rules. Unfortunately, this form of learning becomes quickly routinized and divorced from meaning, thereby leading to the kind of results we all know too well—students who can mindlessly apply rules without having any understanding of what they are doing, and who are unable to use what they have learned to solve any complex or novel problems. But when a teacher shifts the burden of learning to the students, many will resist and seek to restore the mathematics class to what they think it is supposed to be—with the teacher as the presenter of all ideas and the students passive, silent, and working alone, with the emphasis on memorization and rules, rather than on communication and inquiry.

Rome Wasn't Built in a Day: Make Time and Have Patience

It will take time to win some students over to a new view of the mathematics classroom. Experienced teachers know that even though the curriculum is (over)crowded, it contains some dead spots—topics that would never be missed. In fact, the National Council of Teachers of Mathematics *Curriculum and Evaluation Standards for School Mathematics* (1989a) suggests a number of topics that might be skimmed or just mentioned (e.g., the factoring of quadratic expressions with integral coefficients). There is always some resistance to eliminating any topic from the curriculum, but even if topics are not eliminated, teachers can choose to place their emphasis on the *quality* of the problems or exercises to be solved, rather than on the *quantity*. It can be more beneficial to solve a single problem many ways than to solve many problems a single way.

Skillful manipulation of precious instructional time, with an eye toward the relative importance of individual topics in the curriculum, can allow a teacher to find time for more classroom discussions or extended investigations. Of course, it is important that key topics in any course be treated thoroughly and that students be given an adequate opportunity to learn important material. But our goal should be not so much to cover every topic, especially since so much of what we cover is forgotten, but rather to guide and encourage our students to acquire a reasonable level of skill, confidence, and enjoyment in learning and doing mathematics on their own and with their classmates. It is the latter that will ensure not only that students are less likely to shy away from further study of mathematics, but also that the foundation will be laid for their pursuit of lifelong learning that might include mathematics.

Setting Modest Goals: The 10 Percent Solution

Nothing is more frustrating to teachers than to enter class with great expectations for a grand revolution, only to have their hopes dashed by a noncooperative, uninspired group of students. Out of such frustration grow despair and a resistance to try again. Rather than changing everything at once, or expecting instant success, it is wise to start slowly and with modest hopes. Rather than creating a "high-stakes" situation in which neither you nor your students will be comfortable, approach the task with a more experimental attitude. You will probably have to try a new procedure, such

Notes

Notes

as problem solving in small groups or having students keep learning logs, more than once before judging its efficacy. You and your students will have to become accustomed to a novel element in the environment, and it will take time to feel comfortable. Sometimes the innovation will simply not work well in your class. Instead of abandoning it, try to think of ways to learn from the failure and revise your approach. Colleagues of ours in the United Kingdom have a phrase they use to describe their attitude toward classroom innovation of this sort: "Fail fast and fail often!" Their experience, and ours, suggests that if you keep trying and don't become too discouraged when things don't go particularly well, or too elated when they do—and expect to fail at least some of the time—the resulting transformations in the classroom environment and in the quality of interaction and learning can be sensational.

Our colleagues in the United Kingdom also speak of the "10 percent solution" to classroom innovation. They suggest that it is probably possible for an average teacher successfully to alter only about 10 percent of classroom instruction during a school year. This modest goal may serve as a wise target for many teachers in the United States as well.

To change 10 percent of an instructional program requires that teachers alter only one class period every two weeks, five minutes of every class period, or one instructional unit during the year. This small perturbation means that most teachers could find the time to incorporate more communication and inquiry into their secondary mathematics classes. Although the 10 percent goal may seem too modest, and the more adventuresome might wish to go beyond it, all secondary school mathematics teachers will appreciate the tremendous cumulative effects this approach can have over several years.

Possible Entry Points

Assuming you are willing to set modest goals and proceed slowly but persistently to make changes in your mathematics classroom to encourage more communication and inquiry, where should you start? We offer two suggestions: start with "weak spots," or start with "strong examples."

The first suggestion is to start at some point in the curriculum with which you are currently unhappy. All teachers can identify certain weak spots—topics for which they are dissatisfied with their teaching and their students' learning. These weak spots constitute ideal targets for innovation. Since you are already unhappy with the results when teaching in the traditional manner, here is a golden opportunity to try something different. First, give some thought to the desired goals for the topic. Then consider the ways in which traditional instruction fails to promote those goals, and the ways that an innovative approach might be used more successfully. Now you are ready to bring innovation into your classroom. When you begin with a topic you consider a weak spot, you are motivated to change what you have been doing, and the chances are good that you will be able to make some improvement. Successful innovation may require little more than altering the form of some of the problems to make them more open-ended, thereby inviting more inquiry on the part of the students.

Another strategic approach to innovation is to use the great range of published, high-quality instructional prototypes that might be called strong examples. Many professional journals and resource books contain examples of successful lesson prototypes. Articles in journals often suggest ways of treating traditional topics in nontraditional ways. Although journals and books may have actual lessons, many teachers will prefer to adapt the lessons to their own style. Nevertheless, the suggested topics and approaches can serve as a useful starting point for innovation. Presentations at workshops and professional meetings may also contain exemplary lesson material or useful ideas for new approaches to curricular topics. And don't forget to consider some of your favorite problems. Old chestnuts, like the chickens-and-rabbits problem used by Mr. Westerman in Chapter 2, can often serve as starting points. Colleagues are also a useful source of innovative ideas and strong examples. Ask your fellow teachers, especially those whom you respect, what has worked for them, and be prepared to share your successes (and failures) with them. Communication among mathematical colleagues is important to the enhancement of the discourse that you seek to create in your classroom.

A Case in Point: How Mrs. Holmes Got Started

In November of one school year, Mrs. Holmes and her colleague Mr. Jarvis, a fellow mathematics teacher at her high school, attended a conference on mathematics teaching where they participated in a session on problem solving. Mrs. Holmes decided to try one of the problems with her sophomore beginning geometry class. Late one Friday afternoon, 15 minutes before the end of the period, she put the following problem on the overhead projector and asked her students to think about it:

> *Eight people attended a business meeting. If everybody shook hands once with everybody else, how many handshakes were exchanged?*

Immediately, students began shouting out their answers: "64" and "16"! Mrs. Holmes said, "No, no! Don't just guess. Think first!" That stopped the guessing, but the class quickly fell apart into a series of side conversations, many of which were not related to the problem or even to mathematics. Some students began looking at the clock, and others simply stared out the window. When Mrs. Holmes asked if anyone had a solution, she was greeted with silence. Mercifully, the bell soon rang, and the school day was over. She was so upset that she forgot to remind the students of their quiz on Monday.

After school, a disappointed Mrs. Holmes conferred with Mr. Jarvis who just had a similar experience in his algebra class. Although they were both dismayed by their students' response, they decided to think about their

Notes

Notes

experiences over the weekend and to meet at lunch on Monday to develop a fresh approach. Over the weekend, Mrs. Holmes looked over the notes she had taken at the conference. She also read an article on problem solving in a recent issue of *Mathematics Teacher*. At lunch on Monday, she and Mr. Jarvis developed a new approach—one that involved some initial investment of their class time in order to achieve a long-term benefit for their students.

Recovering from the Opening Fiasco: A Strategic Approach

Mrs. Holmes devoted the next geometry period entirely to problem solving. First, she divided the class into random groups of four and had the students turn their desks around so they were facing each other. She gave each student a list of problem-solving strategies that she had learned about at the mathematics conference. The list included such strategies as "Make a table," "Look for a pattern," "Draw a diagram," "Guess and test," "Act it out," "Solve a simpler problem," and "Examine extreme cases." She asked the students to put the list in their notebooks and keep it handy for easy reference during their problem-solving activities. She also gave each student a set of six problems, one of which was the "handshake problem" from the previous class session, and passed out scratch paper to encourage students to write and draw their solutions. She told each group to select three problems and to develop a group solution for each. They could use any of the ideas on the strategy sheet to solve the problems. She let them know that about 20 minutes before the end of the period they would stop and talk about the problems. As they worked, she walked around, conferred with students, and answered questions—but she would not solve the problems for the students.

The class got to work. It was noisy. There was arguing and laughter. Mrs. Holmes moved around, clarifying questions, giving occasional hints, and trying to keep the students focused on the task. Students would come up with an answer and demand to know if it was right. Although it was very hard for her, she would not tell them. She challenged them to find a way to convince themselves and her that it was correct. Some of the students found the experience very frustrating, as if they resented being asked to think, but most of the students seemed to enjoy the activity once they got settled into it. At the end of the working time, each group had correctly solved at least one problem. Some groups had solved all three.

When it came time to discuss the problems, she asked for volunteers. Kuo solved the handshake problem by drawing a picture (see Figure 4-1 below). She then asked if anybody else had solved that problem in a different manner. Melanie presented her solution, which involved making a table (see Figure 4-2 below). John said that his group had wanted to act out the problem, but there weren't enough people in the group. So Mrs. Holmes had eight students come up and go through all the handshakes. Suddenly, the period was over. As the students were leaving, Mrs.

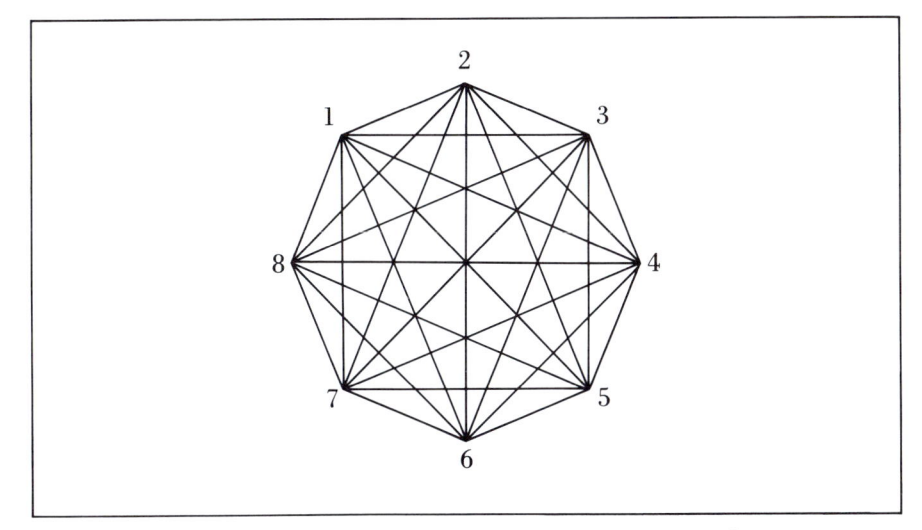

Figure 4-1. Kuo's Solution to the Handshake Problem.

Holmes reminded them to keep their work papers in their notebooks. She was feeling much better about her problem-solving adventure than she had the previous Friday, and she felt that the time she and Mr. Jarvis had spent in mapping out a strategy had been very worthwhile.

Over the next two weeks, using bits and pieces of leftover class time, Mrs. Holmes and her students finished discussing the five additional problems. During this time she reviewed some of the strategies that were less

Person	Shaker	Number
1	2, 3, 4, 5, 6, 7, 8	7
2	3, 4, 5, 6, 7, 8	6
3	4, 5, 6, 7, 8	5
4	5, 6, 7, 8	4
5	6, 7, 8	3
6	7, 8	2
7	8	1
8		0
	Total Shakes	28

Figure 4-2. Melanie's Solution to the Handshake Problem.

GETTING STARTED

Notes

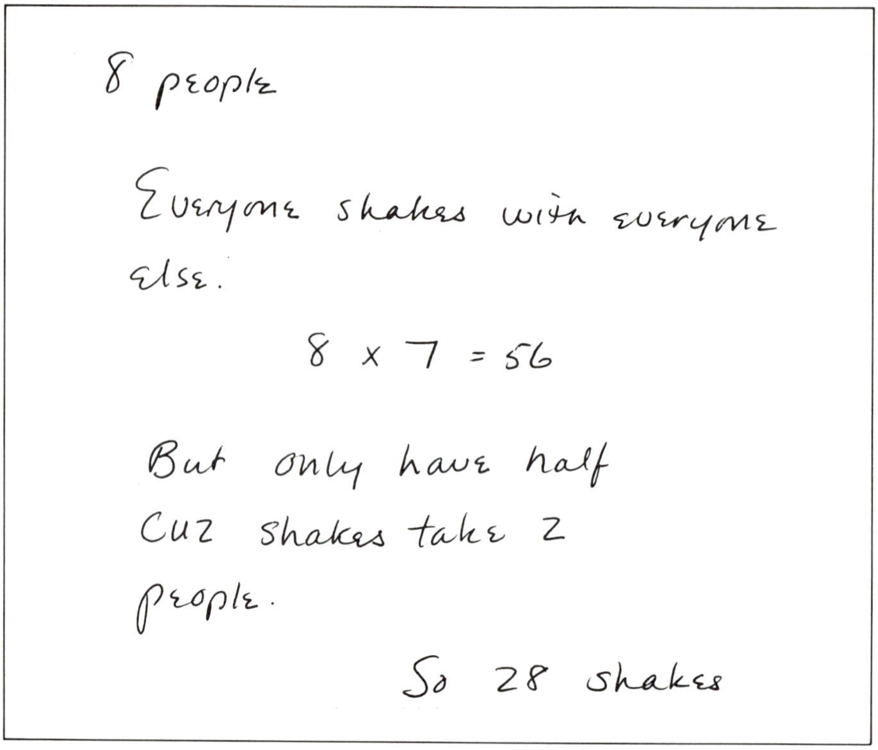

> 8 people
>
> Everyone shakes with everyone else.
>
> 8 × 7 = 56
>
> But only have half cuz shakes take 2 people.
>
> So 28 shakes

Figure 4-3. Tyrone's Solution to the Handshake Problem.

familiar to her students, like reminding them to "Examine extreme cases." The class kept a record of the problem-solving strategies that were used and discussed. Students noticed that some problems could be solved by several strategies, and that certain strategies were used more frequently than others. They began to develop an effective repertoire of strategies that they could use when solving a new problem. If they got stuck, they would consult their own printed list of strategies or the one that Mrs. Holmes had enlarged as a poster and hung in the classroom. Mrs. Holmes was surprised at how long it took to discuss the problems thoroughly, but she could see that it was time well spent. Her students were animated about their problem-solving experiences, and they appeared eager to share solutions and approaches. Although Mrs. Holmes was somewhat dismayed that the class had fallen behind by a day in the district pacing schedule, she decided that the enthusiasm of her students more than compensated for the slight delay.

Mr. Jarvis had also had a successful problem-solving experience with his elementary algebra class. Now the teachers' dilemma was finding a suitable way to keep the momentum going. Since they were both reluctant to spend another entire class period on problem solving, they decided to start a problem-of-the-week contest. They scoured professional journals, puzzle books, and other sources to find interesting problems. It was not always possible to find a problem directly related to the course content being studied in any given week, but Mrs. Holmes tried to find problems with a geometric flavor, and Mr. Jarvis searched for those that might involve some algebra. Each Monday, the students would find a new problem-of-the-week on the side board of their classroom. Students were allowed, actually encouraged, to work collaboratively to solve each problem. Each Friday, students had an opportunity to present and discuss possible solutions.

Mrs. Holmes and Mr. Jarvis both tried to integrate their ideas on problem solving as much as possible into the course content they were teaching. They would refer to and use various strategies when solving example problems in class, and they encouraged students to use the strategies whenever they wanted on their homework exercises. To stimulate discussion about the strategies, they would take some of the routine problems in the textbook and pose them in a more open-ended way, as Mr. Ruiz did with the problem in Figure 3-1 and Mrs. Garcia did with her "Tell me everything you know about a circle" question.

On the Dotted Line: Mrs. Holmes Helps Her Students Write About Mathematics

In February, Mrs. Holmes decided it was appropriate to put the following nonroutine problem on a chapter test:

> *Highways need to be constructed between eight cities so that there is a highway directly connecting every pair of cities. How many highways are needed?*

To motivate the students, she decided to make it an extra-credit problem, worth an extra ten points. She told the students that, in order to get full credit, they would have to show all their work and explain their solution fully. The problem was similar to the handshake problem that had been solved in class, and it was solvable in at least two different ways, using strategies that were now part of the class repertoire. When Mrs. Holmes graded the tests, she found that about half the class attempted the extra-credit problem, and most of these students had the correct numerical answer, but only three students had provided a complete, correct explanation. Most students had simply written down the answer (usually unlabeled). A few had shown some of their arithmetic, and a few others had shown diagrams that they had attempted to draw. Mrs. Holmes discussed her findings with Mr. Jarvis, who had not yet tried putting a nonroutine problem on one of his tests. Together, they decided that although the students had become quite comfortable in discussing mathematical problems and solution strategies, their oral fluency might not automatically transfer to written work. Their students were simply unaccustomed to writing about mathematical ideas.

Mrs. Holmes discussed her experience with a fellow mathematics teacher who had recently attended a district-sponsored workshop on "Writing Across the Curriculum." Mrs. Holmes borrowed the teacher's notes

Notes

Notes

from the workshop in order to develop an approach to helping her students explain their mathematical thinking in writing.

As she had done when she first tried the problem-solving activities, Mrs. Holmes divided the class into groups, each containing four students. She distributed two problems and told each group to choose one to discuss and solve. After a reasonable amount of time, she asked the students to return to their individual desks and write out the solution to the problem. A few students greatly resisted this activity—imagine writing in a math class! Especially resistant were the students for whom English was a second language. She found it helpful to group the limited-English students together when they shared the same native language. For example, she paired two Vietnamese students who spoke the same Chinese dialect. At first, they spoke mostly in Chinese, but gradually they communicated more in English.

Despite some difficulties and resistance, Mrs. Holmes persisted, and finally all the students turned in their written solutions. When she evaluated their written work, Mrs. Holmes made many comments on their papers. She was very generous in her praise of students' attempts to explain their reasoning, although she did try to make comments that would help them to do a better job next time. Over the next few weeks, she repeated this activity a few more times and found that, with additional practice, the students were less resistant, and that it took less time to complete the activity. During the remainder of the year, as Mrs. Holmes continued to put writing problems on her tests, she noticed that the quality of the students' solutions and the completeness of their explanations gradually improved. By the end of the course, most of her students had learned to explain their ideas and approaches in writing, and to identify and clearly label their answers. Moreover, most of the students came to view written and oral communication as a natural part of their mathematics class.

Nothing Succeeds Like Success

It was not only the students who learned something about themselves and mathematical problem solving. By the end of the course, Mrs. Holmes had developed the courage (which she was quite sure she would never have) to attack a brand-new problem in front of the class, modeling her thinking processes as she worked toward the solution. Occasionally, she was unable to solve a problem completely in one day, and it became a challenge problem for everyone to work on overnight. Her students had learned that not all problems could or should be solved in five minutes or less, and that it was all right to work on a problem over a period of time, so they were not troubled by these problem-solving episodes.

Mrs. Holmes's interest in problem solving also affected her colleagues, as word of her innovations spread around the school. Other teachers began to give her problems to solve. In fact, one of her favorite problems came from the school custodian. In December, she and Mr. Jarvis had occasionally met to work on problems at lunch. Gradually, a colleague or two began to join them. By March, an informal problem-solving group evolved which met at lunch once a week. Problems were shared, discussed, and argued over. Eventually, the discussions moved beyond the problems to considering issues in the teaching of mathematics. As her colleagues learned of the successes of her students and those of Mr. Jarvis, they wanted to get their own students involved. They met with the principal and planned a mathematics department in-service program on problem solving and writing in mathematics for the beginning of the next school year. Although many of Mrs. Holmes's colleagues were somewhat skeptical about investing a lot of time in correcting students' written work, she was able to convince them that it would be time well spent, and that students' improved communication skills were worth the investment.

Looking Back: Mrs. Holmes Reflects on the Year

Was the extra time and energy worth it? Mrs. Holmes conducted an anonymous survey and found that the students were quite positive about the problem-solving experiences they had had during the year. They felt more confident about themselves as problem solvers and agreed that the problem solving was an enjoyable and worthwhile addition to the class. Mrs. Holmes was pleased that more of her students than usual elected to continue in the college-preparatory mathematics sequence at her school. In her overall evaluation, Mrs. Holmes decided that she was quite satisfied with her approach to problem solving and decided to incorporate it in all her classes the next school year.

As she reflected on the year's experience, Mrs. Holmes also realized that she had learned to use her class time much more efficiently than before. If there were a spare minute or two at the end of the period, she would quickly give the class a problem to solve. However, even with her efficient use of time, she found that she had to sacrifice a small amount of the geometry content toward the end of the year. She covered all of the necessary material, but skipped some of the repetitious proofs during second-semester geometry. She decided that her students had sufficient knowledge of geometric proofs, and that it was more important for them to learn a broader range of thinking and analytic skills that would serve them well in future mathematics classes—and in life. Given the enthusiastic response of her students, their increased willingness to grapple with hard problems and discuss mathematical ideas, and the evidence that they had learned the important geometry course content very well, Mrs. Holmes was quite pleased with the results of the year.

Lessons to Be Learned: Further Reflections on Mrs. Holmes's Experience

Mrs. Holmes's story demonstrates a number of important points. First, it really helps to have a colleague. She clearly benefited from her conversations with Mr. Jarvis, in which he not only provided support and encour-

Notes

agement, but also made substantive suggestions regarding problems, activities, and approaches to be used with students. Mrs. Holmes was determined to teach her students problem solving, but she was also realistic. She did not try to change everything about her teaching. Most of what she did remained very similar to what she had done for years, but she managed to encourage more inquiry and communication in her classroom than she ever had before. Mrs. Holmes was also patient and flexible. She did not expect her students to change immediately, and she was willing to encourage her students during their slow, steady journey toward their goal. When things did not go well—when she met resistance or failure—she was willing to step back, carefully evaluate the situation, and seek collegial help before reformulating her approach in order to overcome the obstacles. She did not hesitate to experiment, and was open to ideas from many sources, including professional journals, meetings, and her colleagues, but she also relied on strong examples and well-known problems, such as the handshake problem. She experimented with a variety of approaches—group work, whole class discussions, open-ended investigations, writing assignments—until she found the right combination for herself and her students. Mrs. Holmes made time for these activities and learned to use her time and that of her students efficiently and effectively to derive maximum benefit for her students.

Mrs. Holmes's story illustrates that a mathematics teacher can successfully modify instruction to foster increased student inquiry and communication—making the mathematics classroom a place for thinking through mathematics.

References

Appel, K., and W. Haken. The Solution of the Four-Color-Map Problem. *Scientific American* 237:108-121.

Bishop, A. J. 1988. *Mathematical Enculturation: A Cultural Perspective on Mathematics Education*. Dordrecht: Kluwer.

Brown, S., and M. Walter. 1983. *The Art of Problem Posing*. Hillsdale, New Jersey: Lawrence Erlbaum.

Kieran, C. 1981. Concepts Associated With the Equality Symbol. *Educational Studies in Mathematics* 12:317-326.

Lakatos, I. 1976. *Proofs and Refutations: The Logic of Mathematical Discovery*. New York: Cambridge University Press.

Mathematical Sciences Education Board. 1989. *Everybody Counts: A Report to the Nation on the Future of Mathematics Education*. Washington, D.C.: National Academy Press.

National Council of Teachers of Mathematics. 1989(a). *Curriculum and Evaluation Standards for School Mathematics*. Reston, Virginia: Author.

National Council of Teachers of Mathematics. 1989(b). *Professional Standards for Teaching Mathematics*. Working draft. Reston, Virginia: Author.

Peterson, I. 1988. *The Mathematical Tourist*. New York: W. H. Freeman and Co.

Pollak, H. O. 1970. Applications of Mathematics. In E.G. Begle, ed., *Mathematics Education*, pp. 311–334. Chicago, Illinois: The National Society for the Study of Education.

Polya, G. 1981. *Mathematical Discovery*. New York: John Wiley & Sons.

Resnick, L. 1987. *Education and Learning to Think*. Washington, D.C.: National Academy Press.

Resnick, L., and L. E. Klopfer, eds. 1989. *Toward the Thinking Curriculum: Current Cognitive Research*. 1989 Yearbook of the Association for Supervision and Curriculum Development. Alexandria, Virginia: ASCD.

Senk, S. 1985. How Well Do Students Write Geometry Proofs? *Mathematics Teacher* 78:448-456.

Senk, S., and Z. Usiskin. 1983. Geometry Proof-Writing: A New View of Sex Differences in Mathematics Ability. *American Journal of Education* 91:187-201.

Steen, L. A. 1988. The Science of Patterns. *Science* 240:611-616.

Tymoczko, T., ed., 1985. *New Directions in the Philosophy of Mathematics*. Boston: Birkhauser.